FORCES AND MOMENTS - 2

It is the first book in a large and special series of books, dedicated to motorsport in general; it will cover aerodynamics, suspension, engines, dynamics, etc. ..... Everything you need to learn how to design a full car.

The aim of this series is also to say that I would like to teach again in a university.

I hope that this series will be a success and that I will be able to transmit all my knowledge and all my experience.

@TimoteoBriet

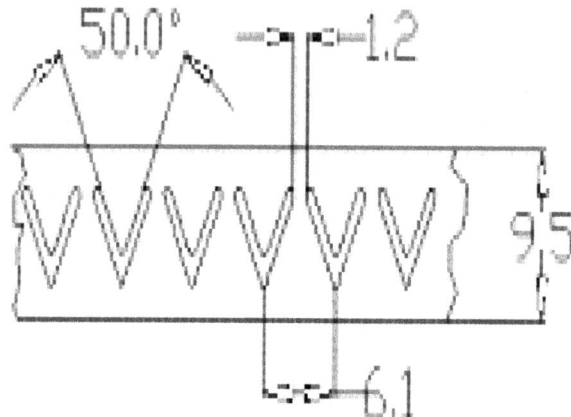

These type of vortex generator, can be used in aerogenerators blades; in this case:

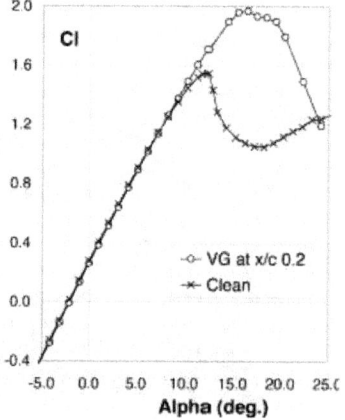

Effects of Vortex Generators on the performance of DU 97-W-300 airfoil

Increasing efficiency turbine blades:

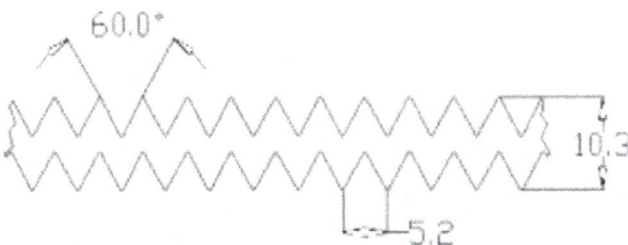

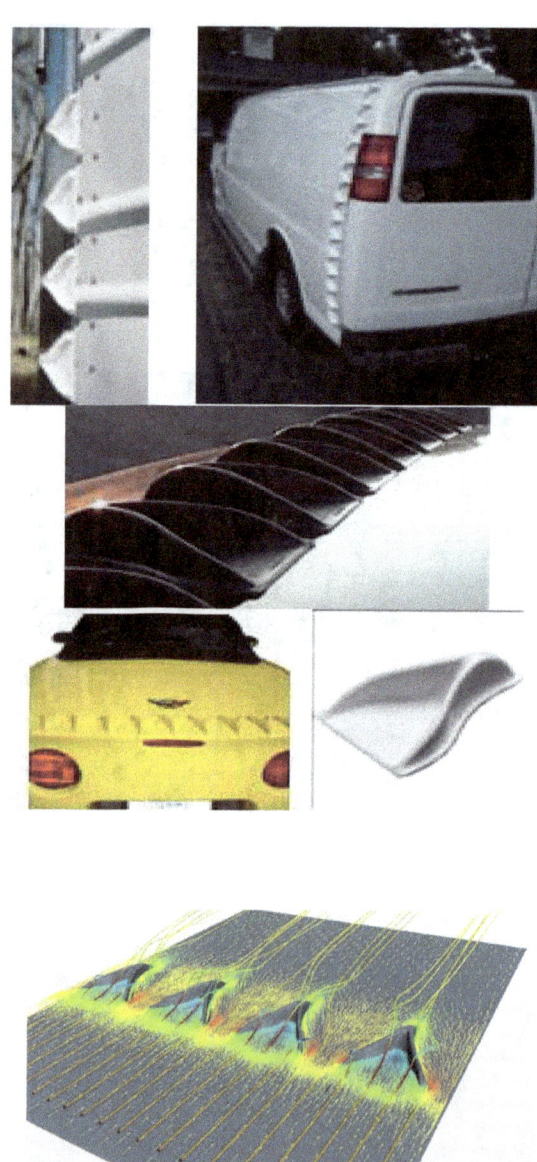

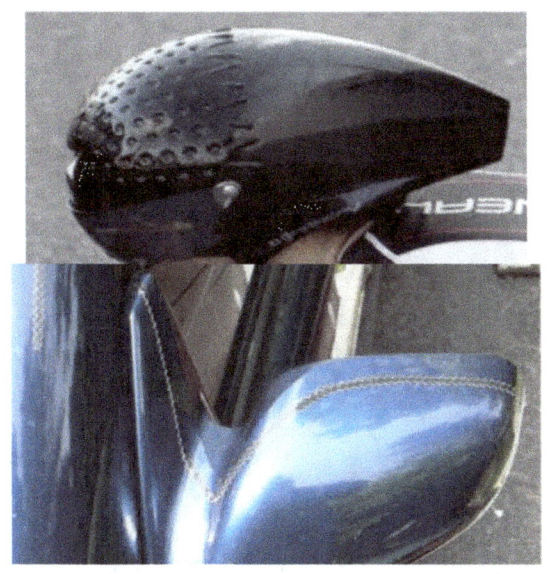

We can see other turbulator designs:

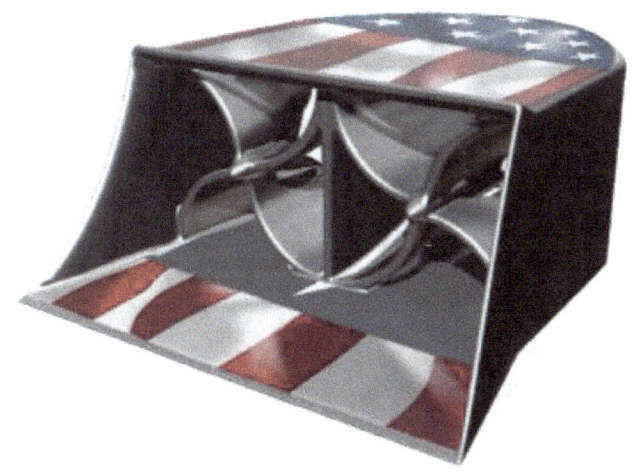

- Divide and win:

We can divide a turbulence, so that the drag generated by this new turbulence is smaller and the turbulence has less energy than the one generated by the original turbulence.

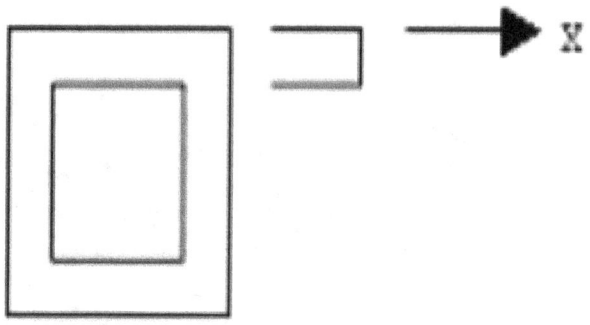

If we denote "Anc" as the width of the box:
x = 0.06 Anc
And "L" as the plate's depth: L = 0.36 Anc

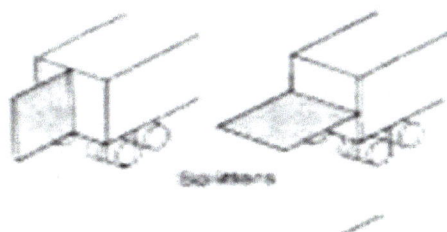

Splitters

Cavities

U.S. Patent Jul. 22, 1986 Sheet 1 of 3 4,601,508

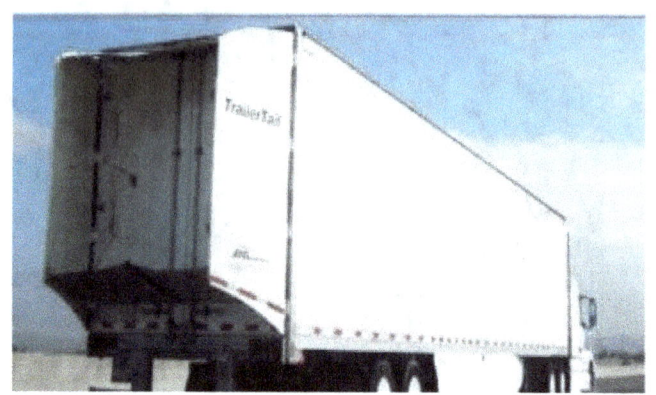

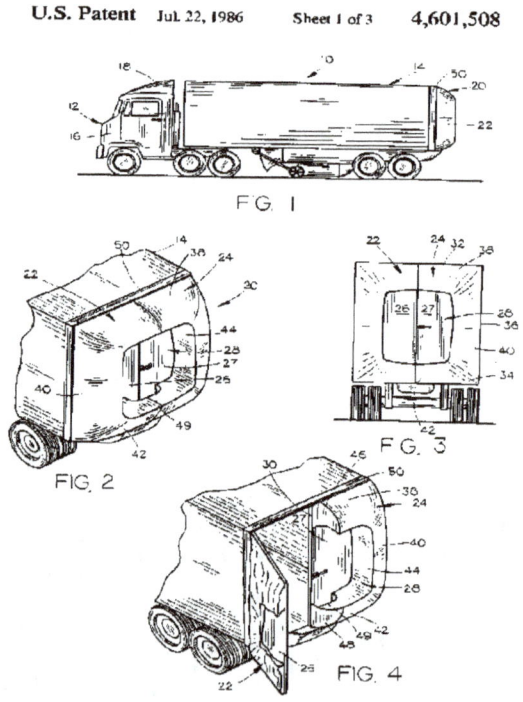

FIG. 1

FIG. 2

FIG. 3

FIG. 4

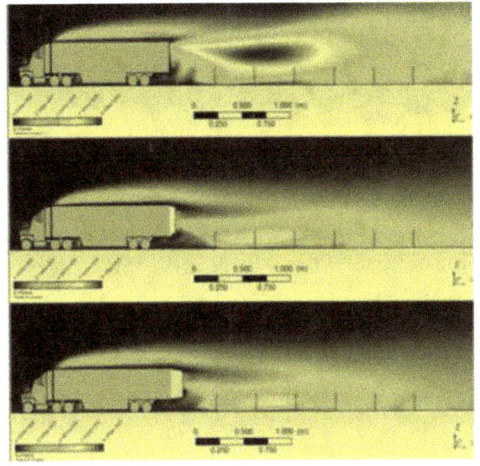

Remembering previously
seen profiles:

The goal is the same: at each "step" we generate a depression which in turn sucks the flow keeping it attached to the surface:

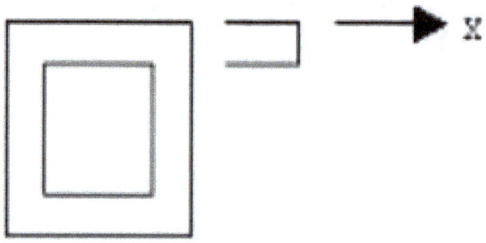

Following the principle of dividing turbulences to achieve less total energy, we find the following system: it divides the turbulence generated behind the caravan:

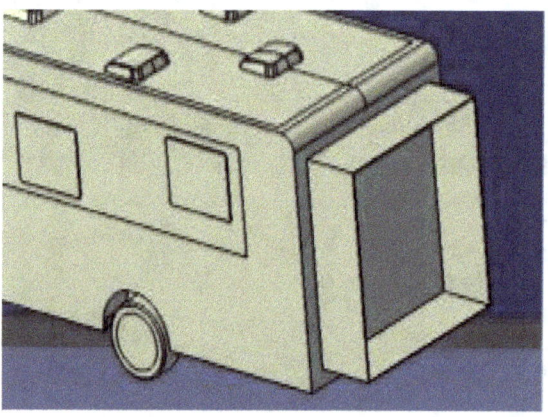

The rear turbulence (wake) is much larger in the case of not having these extensions:

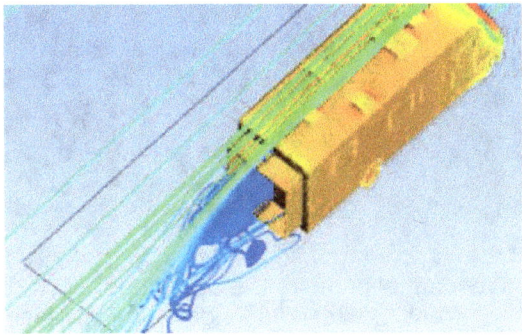

The goal, as always, is to make smaller the virtual geometry.

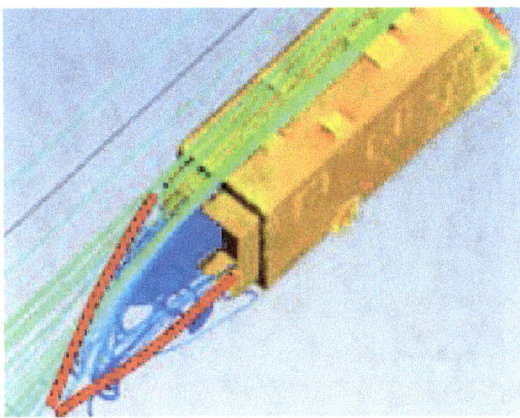

The gap between the truck and the trailer can also be divided:

We obtain a reduction in the total resistance:

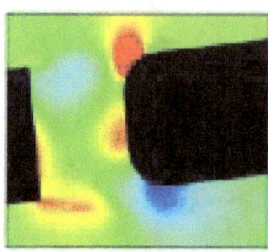 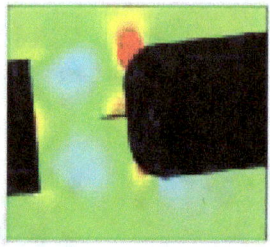

We can also help the reduction of turbulence from the front, extending sidewalls increasing the effectiveness of the method:

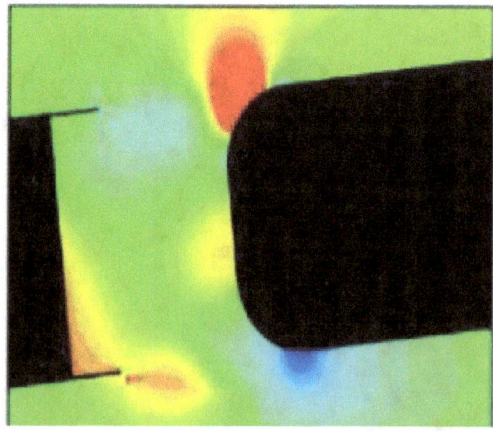

- Rounding edges:

We have seen the importance of rounding sharp edges; let's look at a graph that shows the drag coefficient as a function of the edge radius and Reynolds number:

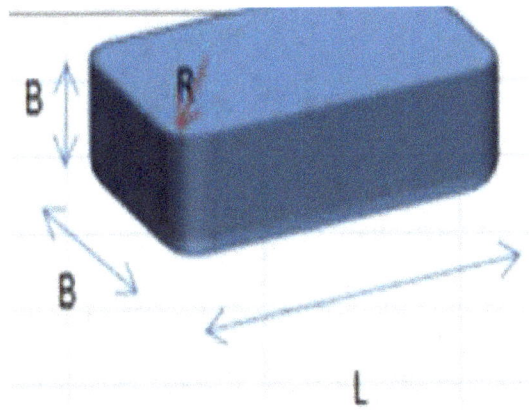

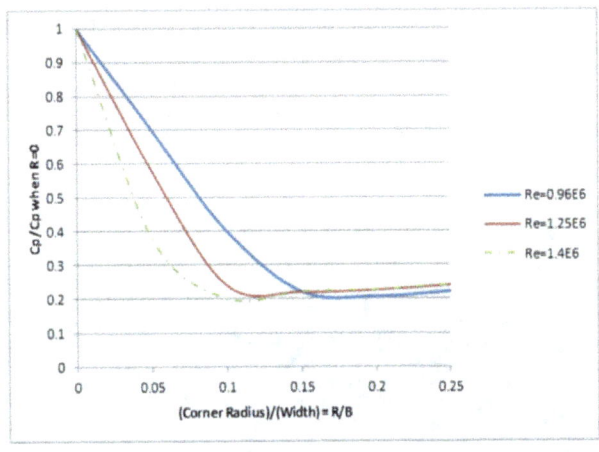

- Refining the rear:

A generic and very efficient method is to sharpen the rear; this is a basic idea but at least in the case of trucks it is difficult for many practical reasons. The dark area is the truck and the light green colored lines are the system to be placed in the rear ("VS" top view, "VL" side view):

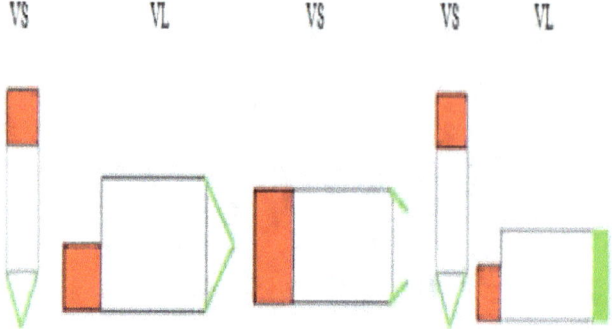

Obviously, the goal is completely soften the rear trying to finish in a point, but as mentioned above in the case of trucks or even cars this is not possible. We tend to approach it as much as possible:

We see the effect of this reduction in the rear:

Minimizing the low pressure is better achieved in this system than without as we can see in the images below:

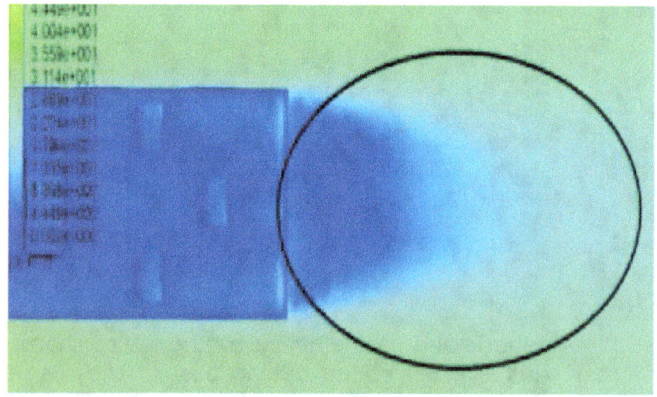

The goal, as always, is to make smaller the virtual geometry:

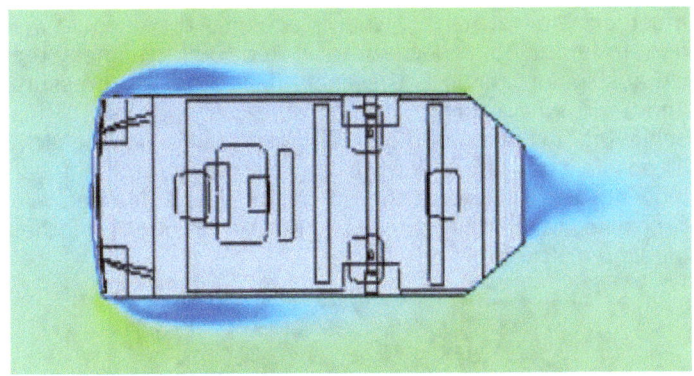

- Filling artificially the vacuum in the rear:

For cycling helmets it is something already used:

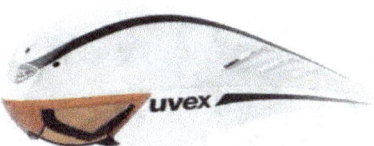

Red Bull uses a system that is combination of these systems:

On one hand we have the gas or cooling air that must go "somewhere"; what place out there could we find in order to reduce drag... Red Bull channels the exhaust gases towards the back, filling the low-pressure and "virtually tuning" the profile of the car. Thus achieving a reduction in drag and increasing top speed; it is possible to direct the air to the top of the diffuser blowing air to increase the efficiency of the diffuser. We can also take this same air intake through another pipeline or from any other site:

Depending on the team, the options are diverse; the air entering the sidepods and responsible for cooling the powertrain can also be used to fill the rear depression:

We can also fill this depression, physically as we saw; but in racing it is not useful:

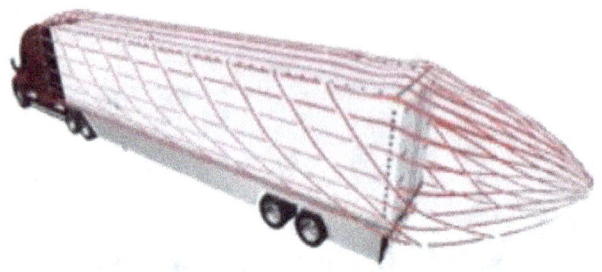

- Flow dividers on a surface:

It's logical to think that a flow in contact with a surface runs along this surface without moving laterally or oscillating; This is not really so: the flow moves laterally losing energy, which ultimately increases resistance power; This method prevents lateral movement by incorporating "barriers":

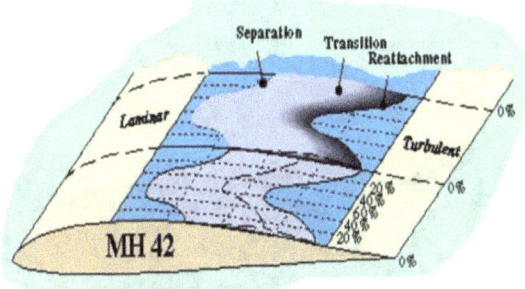

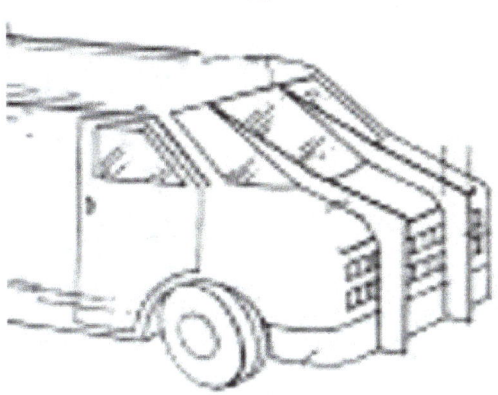

- Other methods:

We will now take a look at other methods or systems that help reduce resistance but whose classification is not too clear; for example, placed in the bottom of the vehicle a part that provides a reduction in drag. Due to its location, it also reduces drag caused by the wheel rotation:

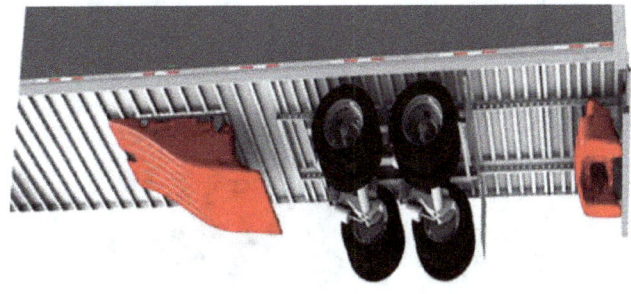

What it is actually occurring when incorporating the device is that the bottom section is reduced thus speed increases; this eliminates the rear depression quicker, reducing drag.

- Softening the rear edges:

Pay attention to the fact that rounded rear edges cause the air flowing over the truck and the sides, enter "quickly" in depression rear; This, in itself, causes a reduction of drag. This smoothing curve "helps" the air to flow into the low pressure area.

The goal, as always, is to make smaller the virtual geometry.
We can find a type of this systems; "INFLATABLE!"

- Roof:

This is an area to design carefully as it contributes significantly to the reduction in overall vehicle drag. The most important parameter is the angle of inclination:

Given a design, there will be an angle where drag begins to increase.

Another very important consideration when designing the back of a vehicle is how long the angled shape will be.

An angle of 21° at the back, usually tends to be ideal:

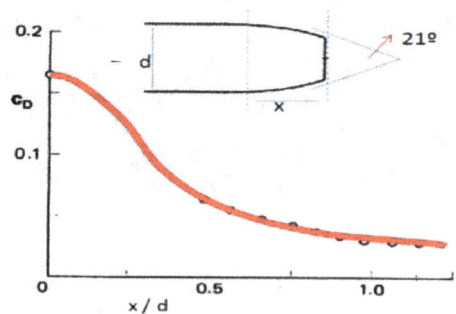

In order to this value is possible to calculate the shape of one body, for attached the air; we suppose that f(x) is a function of shape in two dimensions:

$$f(x) = ax^2 + bx + c \qquad a, b, c \; constants$$

$$f'(x) = \frac{\partial f}{\partial x} = 2ax + b = tg(21^\circ)$$

One equation and two knows; "a" and "b" depend of shape to create.

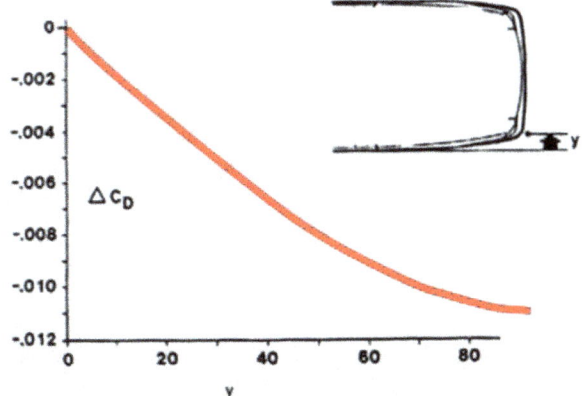

Shape front hood, windshield, side and roof:

The shape of the hood and the windshield are important since they are the parts of initial impact of the air:

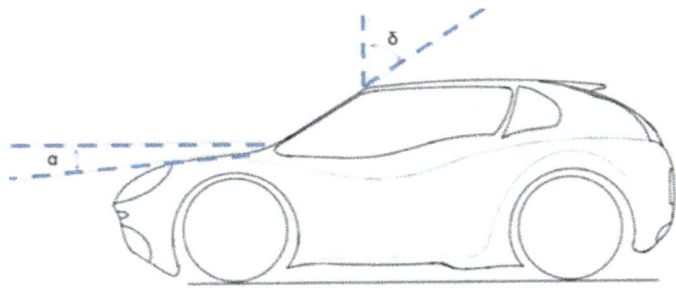

Here we can see the influence of both angles, on the overall vehicle drag:

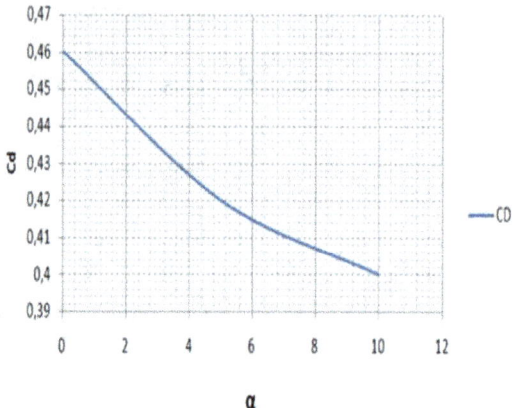

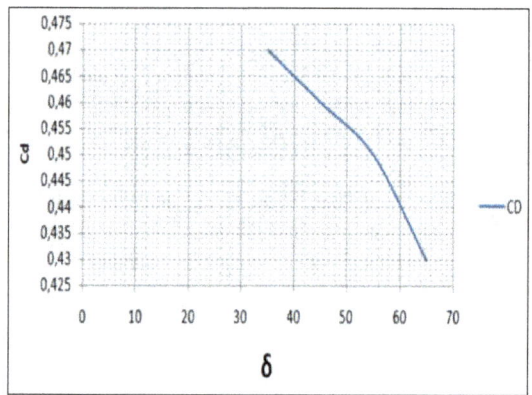

The roof shape and its convexity is also important as it has an influence on drag:

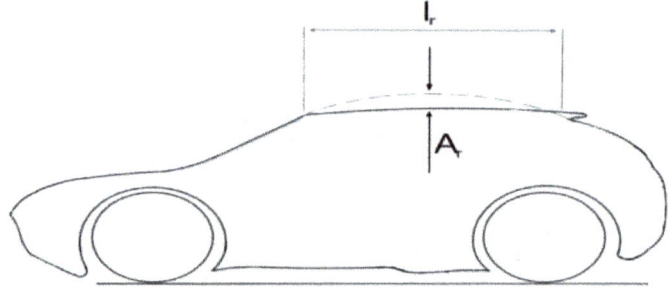

The concavity of the sides is another parameter to consider (horizontal axis = A/l):

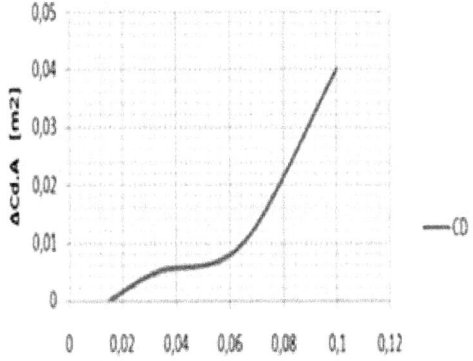

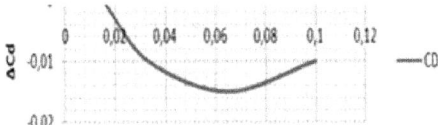

Top view:

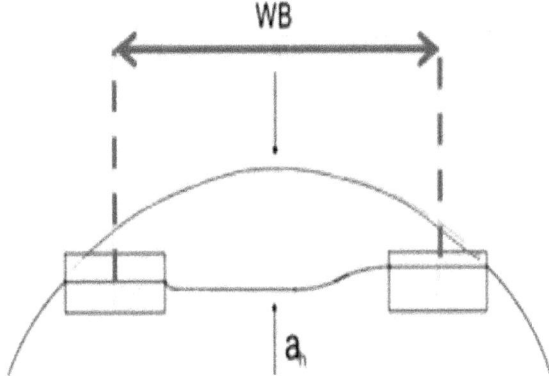

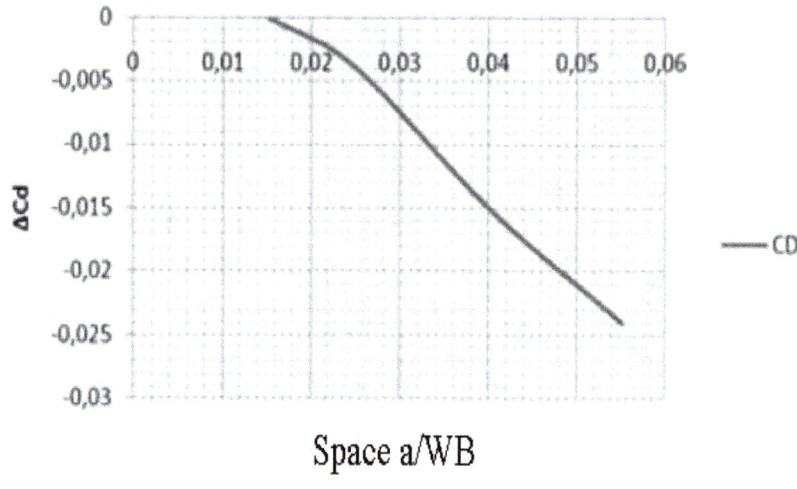

Space a/WB

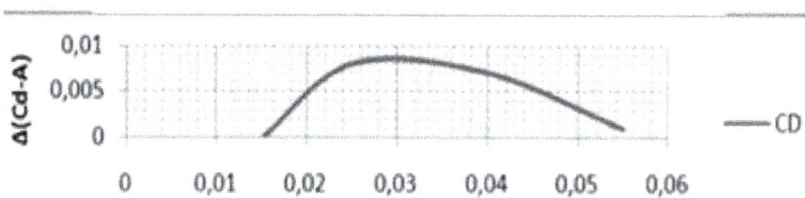

Another graphic-relation, between diffuser angle and shape geometry: the diffuser angle is important, but not is possible to use it always:

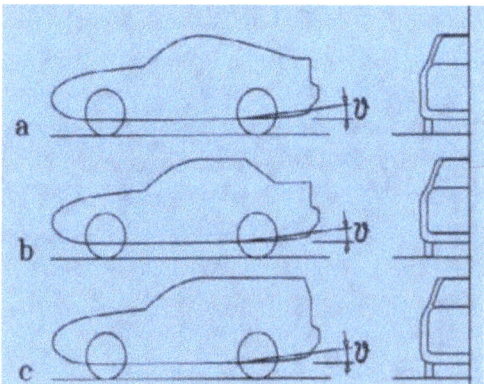

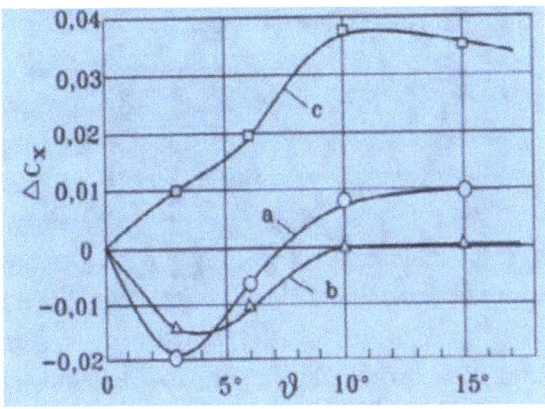

- **Rear ears:**

It is combination of the "divide and conquer" and "filling rear depression" methods. It tries to help the flow move into the low pressure area.

The goal, as always, is to make smaller the virtual geometry.

If you look at the rear and in particular in the vertical ends, we will see a curvature and side walls extending inwardly: they are called ears.

They are an extension of the side walls inwardly, so that the "jump" into depression flow is softer. It is true that this should also be included on top .... but it is not normally practical.

This produces a reduction in drag and thus a reduction in fuel consumption.

Another effect caused by the installation of these rear ears is a reduction in turbulence, which also results in a reduction of drag.

The following image is an exaggeration:

For quantification of its effectiveness, we can use CFD simulation techniques; let's install to ears; one on top and one on the bottom:

We make several trials with different angles (the same angle on top and bottom): 0 °, 5°, 10°, 15°, 20°, 25° and 30°; values "CD" are the values of the drag coefficient; "Cl" is the variations in lift with respect to the case without ears:

Without ears: Cd = 0.35
0°:   Cd=0.3248
    Cl= +0.069
5°:   Cd=0.315
    Cl= +0.014
10°: Cd=0.306
    Cl= +0.032
15°: Cd=0.3115
    Cl= +0.032
20°: Cd=0.3211
    Cl= +0.002
25°: Cd=0.335
    Cl= +0.006

30°: Cd=0.348
Cl= +0.018
35°: Cd=0.36
Cl= -0.001

The angle seems to work best in terms of drag reduction, at 10°.
On the other hand, we must also look at lift; in this case, it doesn't seem to be dangerous the variations of "Cl"

- Pressure releasers:

In the case of high pressure area exists, drawing air in that area to reduce pressure will reduce drag. We can find high pressure in the wheel or on the top of the rear wings.

There are openings, whose function is to reduce drag by injecting air into a depression, but also reducing the pressure in the rear wheel:

They can also act as slits through which air is drawn from inside the "box" of the rear wing or to evacuate the high pressure remaining.

- Change "virtual" geometry; generation of other "not real" geometry:

It is a way of changing the geometry that the air "sees" to reduce drag as the "resultant" body has less drag. Take a sphere: consider that the air goes from left to right in the image; We can make a hole in the left side of the field and inject high velocity air; the resulting geometric body (red line or outer profile) is different from the original sphere and has less drag; It is ultimately, as mentioned above, a "virtually" tuned geometry.

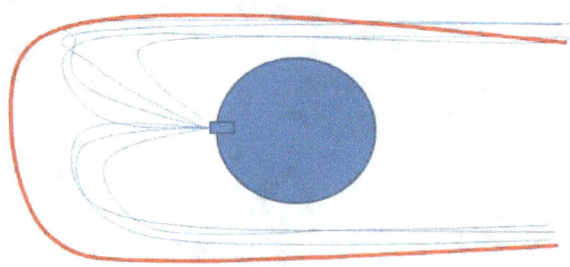

- "Rare" widgets:

Here we just wanted to include a device we saw some time ago placed on the rear wing flap:

Actually we did not know what its use was, although after CFD simulations, we found that the objective was to "reduce" drag and increase lift (in short: improve efficiency).

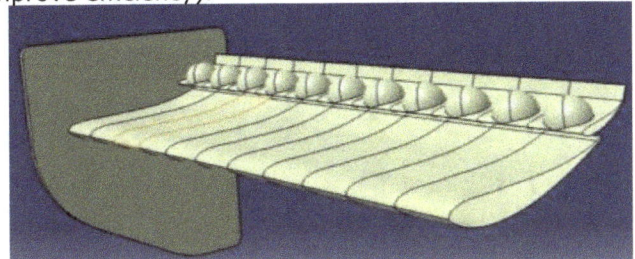

- Adding "more geometry":

We can assume a trail formed by periodic or turbulence; It is possible in some cases to eliminate or mitigate:

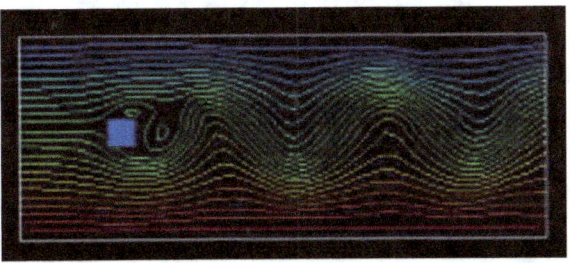

Note:
Let's look at a specific example of study and reduction of drag: The study sought to ascertain the most and least representative areas of a van, to improve those parts where drag is mostly generated and therefore reduce fuel consumption.
The model under consideration is:

The radiator inlet and engine block where included in the CFD study:

The volume mesh was 11 million elements:

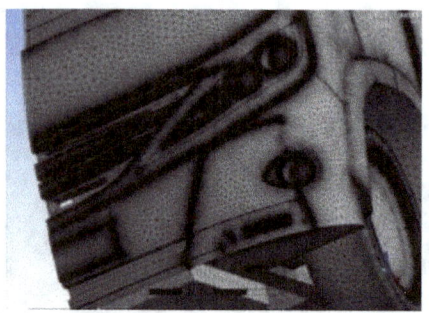

The rear wall is the most important part regarding fuel consumption in terms of the drag generated:

Rear wall	62kg
Windscreen	40.5kg
Front	34kg
Floor	6.6kg
Engine	4.5kg
Laterals	3.3kg
Roof	2.2kg
Cabin roof	-30.8kg

It will be necessary to separate the surface in order to know exactly the transition between the positive and negative drag, in this case negative drag is (arrow):

The van at 100 km/h, needs 40 hp of power (only due to aerodynamic forces).

The most curious thing is:

The cabin roof (windshield-roof junction) produces a negative resistance; this is due to the fact that the air hits the front bouncing and forming a bubble; It helps to have less total resistance.

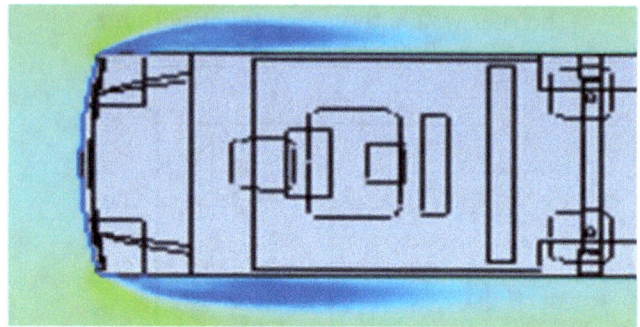

There have been many trials of various systems to reduce fuel consumption.
To reduce the strength of the front, we must focus our full efforts to reduce the low pressure that exists there; this is why it is essential to incorporate air to that area.

To do this, one method is to slow it down just where the edge (A) begins:

In this way the air has less energy to deviate and "falls" literally into the jaws of depression.

We see two graphs of velocity at the edge to appreciate the velocity difference with the addition of certain system (basic model and optimized model):

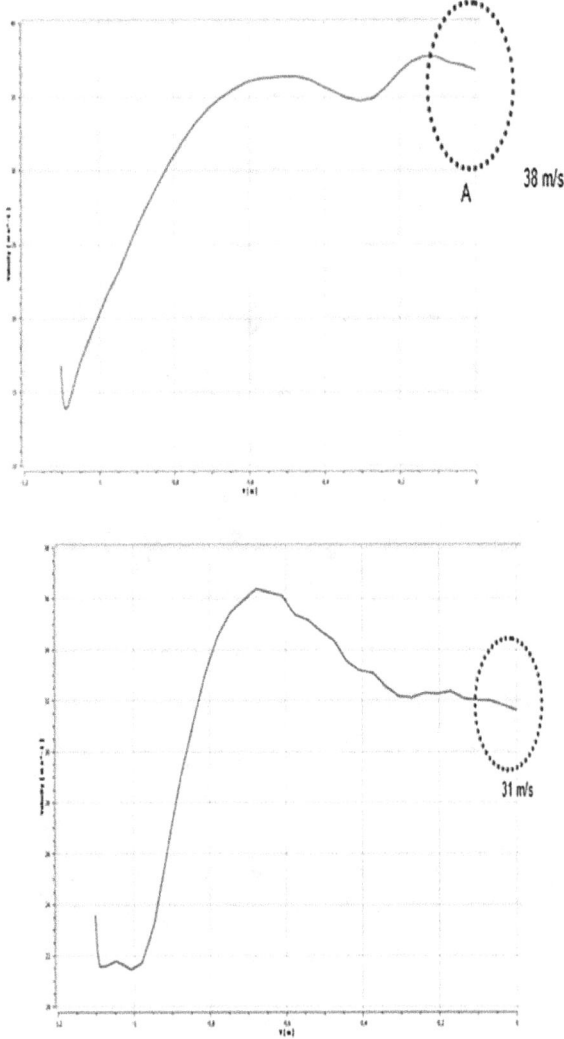

38 m/s

A

31 m/s

Another system for allowing the same are the turbuletors:

Other methods built and tested on the same vehicle as we have seen throughout this book:

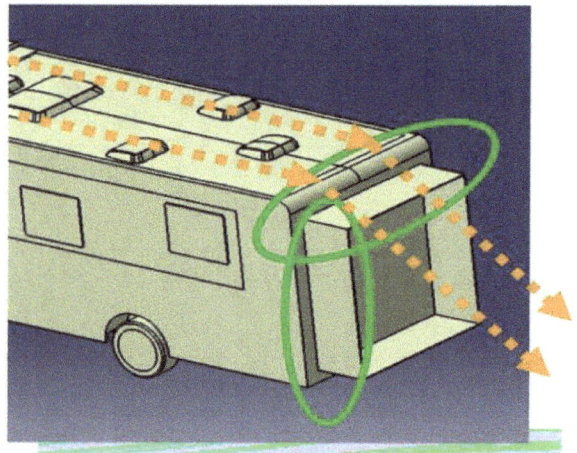

The goal, as always, is to make smaller the virtual geometry.
Circled, these are "artificial" areas which produce a low pressure that suck the flow into the Great Depression.
These are tests, from ideas and more ideas:
Cutting angle: reduction of the rear shape:

Small ears:

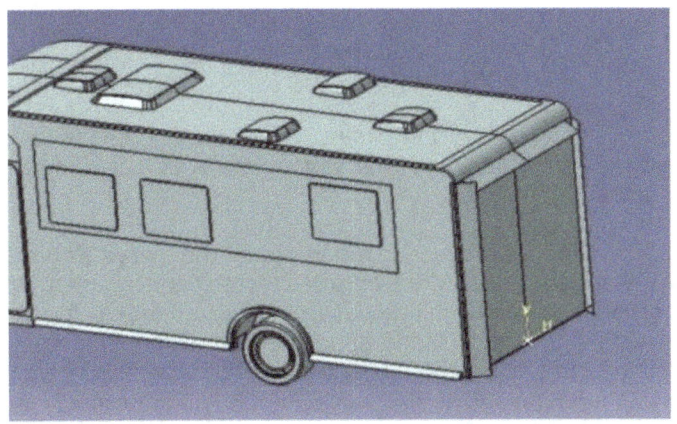

The purpose of including this example here is threefold:

1) An aerodynamics analysis entails two tasks:

Viewing speed streamlines, and pressure fields (colors and pathways).

2) Calculating force values to determine the influence of each part, and therefore know where to act with more emphasis or

interest.

3) Have a simple vehicle model that meets certain "real" rules to determine whether an idea is or is not valid in a "fast and safe" way. Each simulation mentioned above, takes at least 1 full day including post data processing:

   a. Modifying the geometry after eating.
   b. Full meshing is done before dinner.
   c. Simulation program before sleeping and the calculation is launched.
   d. By midmorning the next day, the simulation is over and the information is generated.
   e. We begin another simulation, whose ideas have been pondered the night before; sleeping....

In engineering there is a basic premise and usually acceptable and that works very well: the simplest is usually the best...

For example, consider again the previous van; we want to reduce drag and therefore fuel consumption; perhaps before making complicated changes in the structure or add multiple and other appendages, we can, for example, vary the pitch; we increase the front height 2 cm and go down the back in 2 cm; differences are abysmal relative to drag:

Rear wall	62	kg		Rear wall	47
Windscreen	40.5	kg		Windscreen	39
Front	34	kg		Front	32
Floor	6.6	kg		Floor	3
Engine	4.5	kg		Engine	4.5
Laterals	3.3	kg		Laterals	3
Roof	5	kg		Roof	2.2
Cabin roof	-33	kg		Cabin roof	-30.8

Reductions of around 8% drag.... I said, the easiest and above all "feasible" solutions are perhaps the best and easiest to implement...

In a car, the influence of drag of each part is approximately:

Friction drag	26%
Base drag	35%
Wheel drag	26%
Engine cooling system drag	10%
Induced drag	3%

→ Mercedes concept:

Mercedes-Benz Concept Plug-In Hybrid Transforms To Reduce Drag.

With rear extension, is possible to reduce the drag: from 0.25 to 0.19:

DRAFTING / SLIPSTREAM

We've all heard countless times, the effect slipstream and how important it is in aerodynamics. We know that it is a low pressure generated by the car in front of us, which helps us reduce the power required to advance increasing top speed.

This allows us to prepare the overtaking car ahead of us. But not many people know that the car ahead of us, also benefits from our presence behind....

Let's look at some graphics where we can quantify these mutual effects:

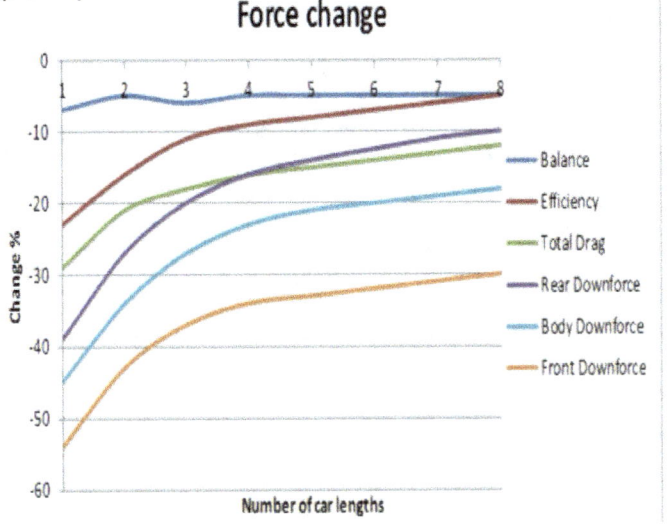

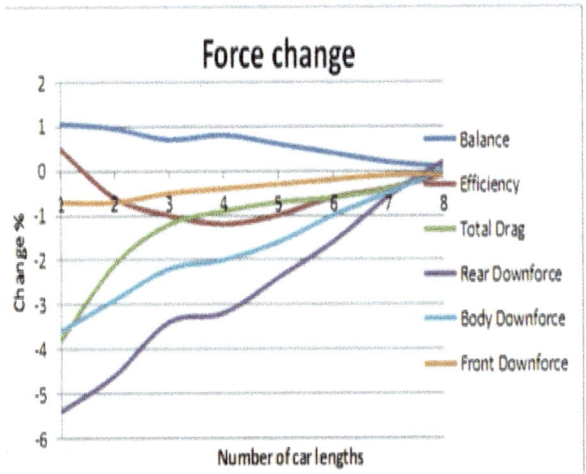

"Car lengths" means how many times the length of the car separates both cars;

If we look closely, we can see what we discussed: both cars benefit, perhaps more the pursuer but also the sought benefits, although not in terms of downforce. We can see this aerodynamic effect in nature: birds on their long journeys, flying neatly behind one another as the air hits them. Swimmers should also take advantage of the "oblique" contrails .If we analyse the resistance generated by a car, we can divided it into two main areas: front and rear. If two cars are one behind the other, the rear resistance of the first car is affected and reduced; so does the strength of the front of the second car; study these mutual effects CFD is complicated because the computation time is enormous. What it is often done is to simulate a single car and its velocity profile downstream, we enter it as initial input velocity profile in another simulation of another car; thus we can observe the mutual interactions.

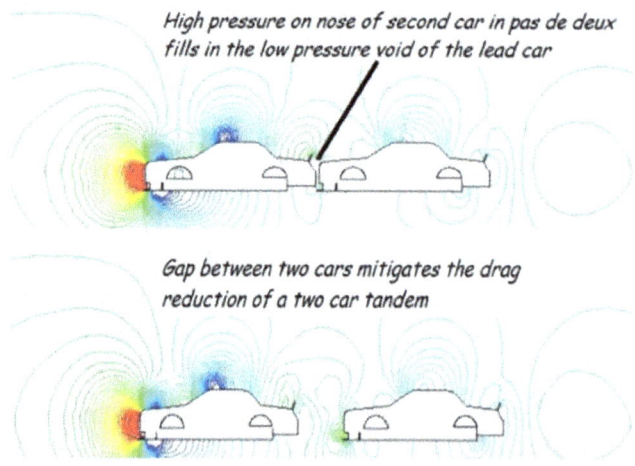

High pressure on nose of second car in pas de deux
fills in the low pressure void of the lead car

Gap between two cars mitigates the drag
reduction of a two car tandem

Let us now look closer at the aerodynamic changes depending on the distances being 2 cars:

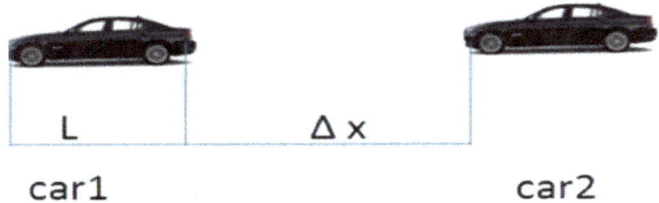

L   Δ x

car1   car2

The graphs below show a variation of front lift coefficient (CLF) and rear coefficient (CLR) as a function of the distance:

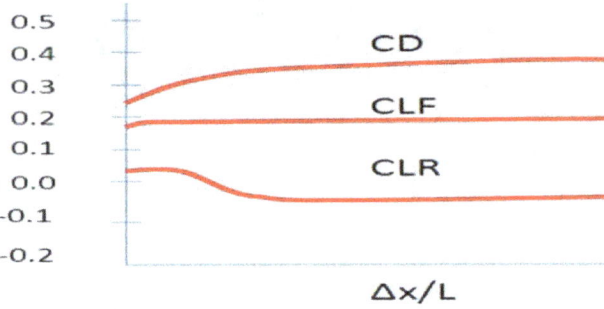

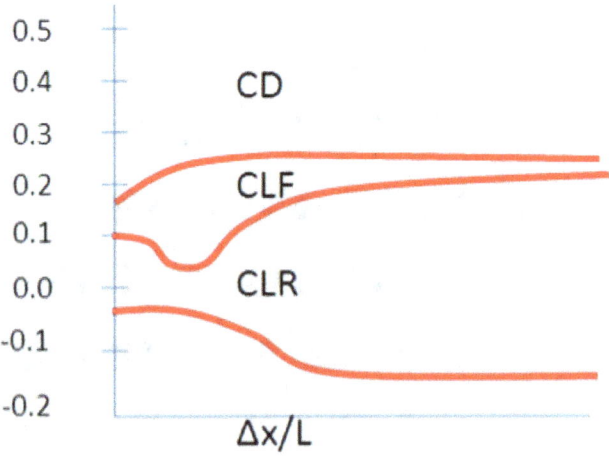

As we see, the variations of the total drag (Cd) and rear lift, are virtually identical; this doesn't happen with the front lift; on the second chase car front lift undergoes major changes and this must be taken into account.

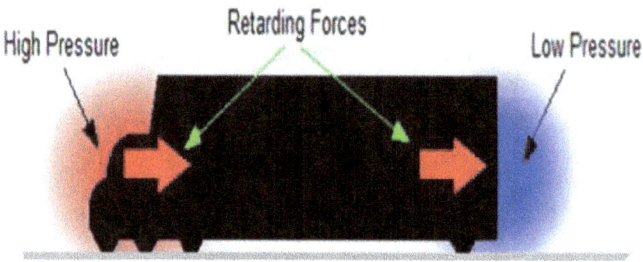

Aerodynamics of a Moving Truck

It is therefore much more efficient for a car to circulate behind other. It is important to know the power requirements of each vehicle forming a line: The following graph shows the energy ratio depending on the number of cars one behind the other:

Car's Energy consumption at 100km/h

Clearly efficiency in terms of fuel consumption and drag generation to circulate in a "row":

The chase car gets turbulent air flow: the flow incident on the chase car is deflected upward (The greater the downforce generated by the preceding car the greater the upflow of air that the chase car will deflect); therefore a pursuing car tends to have understeer, because the car doesn't sufficiently respond to the demand for rotation of the pilot; we have seen in the previous graphs (CLF): as far as possible, the car must be designed to avoid this understeer.

A simulation of three trucks in a line is shown here:

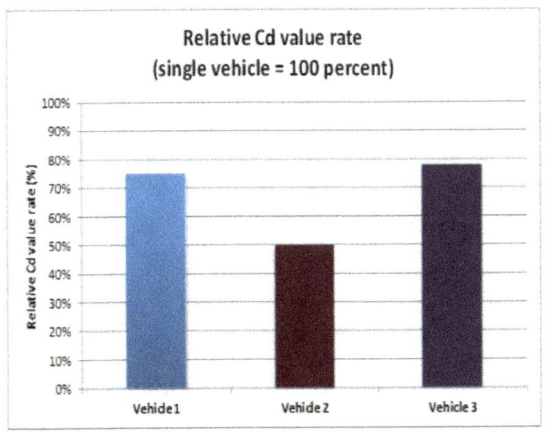

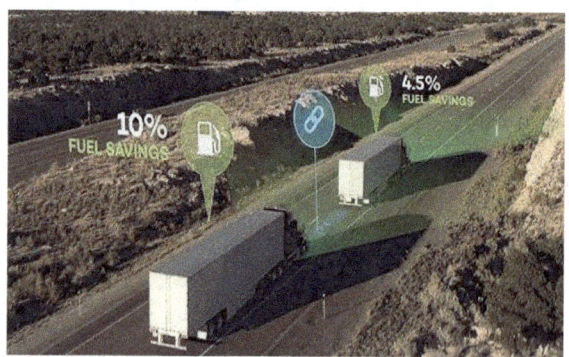

Some graphics about (courtesy of Ecomodder):

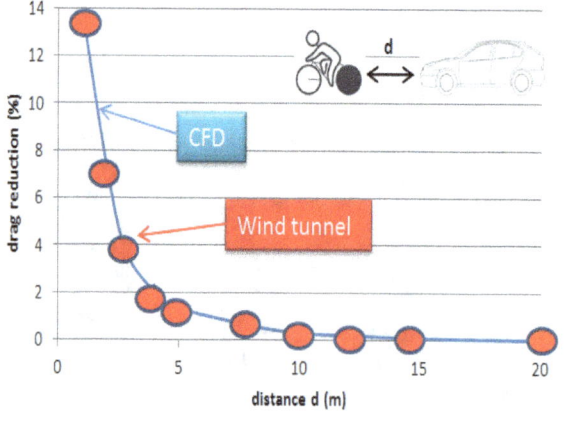

More about that: we analyse the influence of motorcycles, rear bicycle:

At 54 km/h:

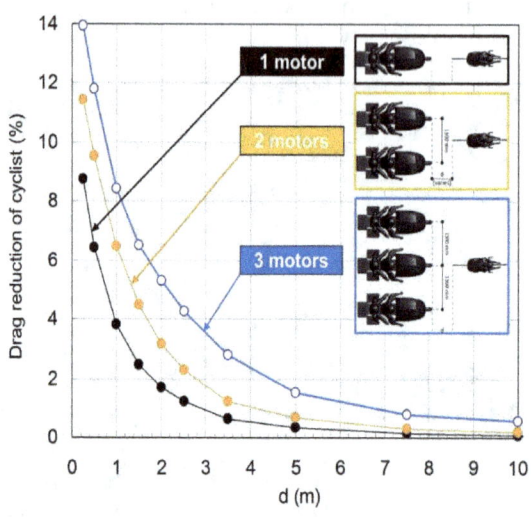

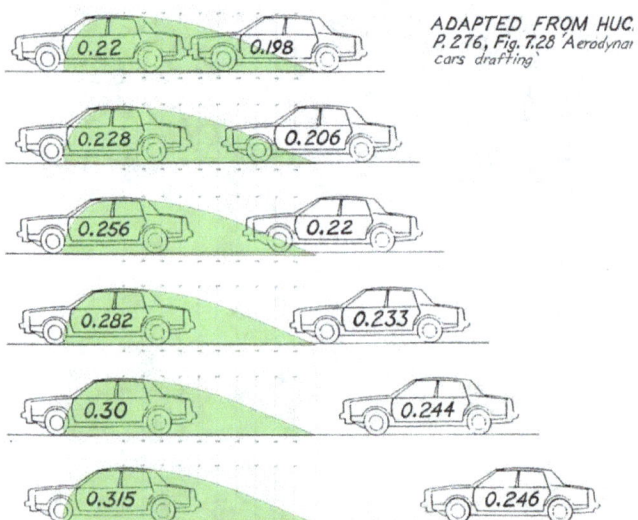

Another of the implications of this mutual interaction in racing cars, is not obvious, but true:

Suppose we change the front wing of a car "standard formula type car"; it is logical to think that the rear wing, like the rest of the car, gets dirtier air (turbulent flow) thus downforce on those parts will be small as well as drag. If we increase the angle of attack of the front wing, the total drag of the vehicle remains the same: we note that even though the front angle ("AV") increases, drag ("SCX") is almost constant:

SCX	FLAPS AV															
	5	6	7	8	9	10	11	12	13	14	15	16	17	18	19	20
5	0,343	0,343	0,350	0,350	0,351	0,351	0,351	0,352	0,352	0,353	0,953	0,953	0,954	0,954	0,954	0,955
6	0,358	0,358	0,358	0,359	0,959	0,960	0,960	0,961	0,961	0,961	0,962	0,962	0,962	0,963	0,963	0,963
7	0,368	0,968														
8	0,380	0,380														
9	0,392	0,392														
10	1,005	1,005														
11	1,033	1,034														
12	1,046	1,047														
13	1,058	1,058														
14	1,063	1,063														
15	1,073	1,073														
16	1,088	1,089														
17	1,097	1,097														
18	1,104	1,105														
19	1,111	1,112														
20	1,118	1,118														

(AILE AR)

What is not as logical or normal, is that if we only change the rear spoiler, the front spoiler also reduces its downforce (remember all parts of a car are interacting with each other).

Question:
Why The DRS is designed to activate the rear spoiler and not the front? So it just said: if the angle of the front spoiler varies, drag, would remain constant, not increasing the top speed.

Suppose the following exercise: let's calculate the drag of the two bodies below a car and a truck:

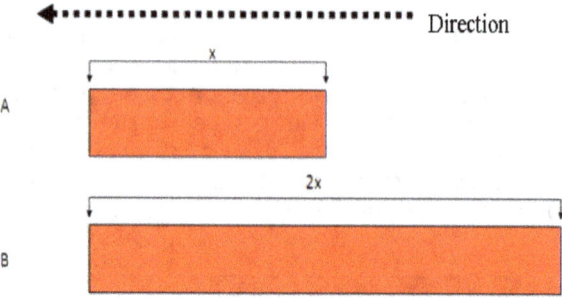

Another factor to consider is the relative position of both vehicles (car and truck). When they are side by side; in this case, the relative size of both vehicles and their distances are fundamental to understand and quantify the interactions between them (the truck is in front):

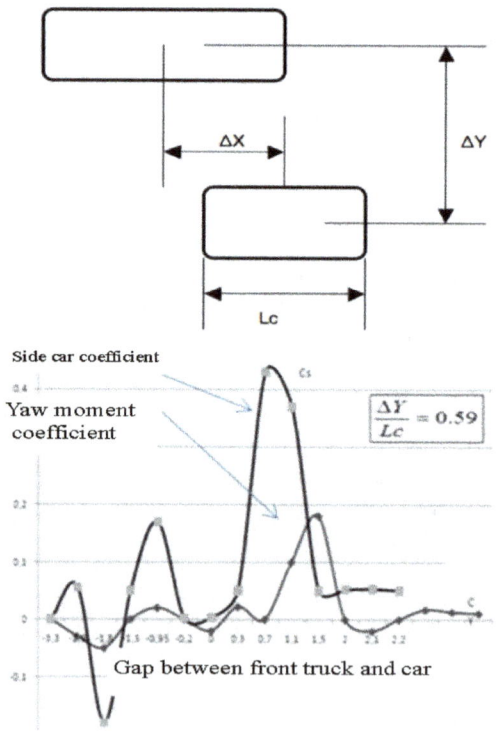

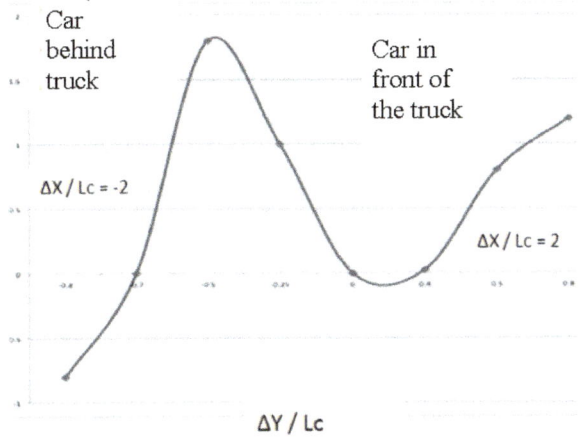

Car behind truck

Car in front of the truck

$\Delta X / Lc = -2$

$\Delta X / Lc = 2$

$\Delta Y / Lc$

Another example:

We suppose that one car want overtaking a truck; show here the trajectory of the car:

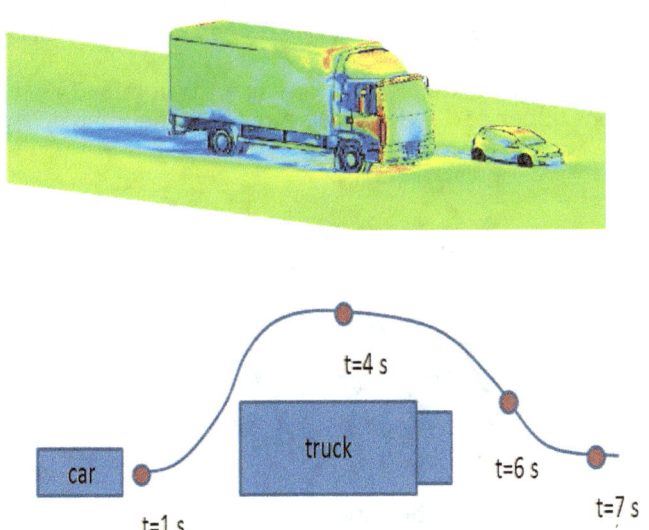

Is possible to know the variations against time, of the drag and lift, of car and truck:

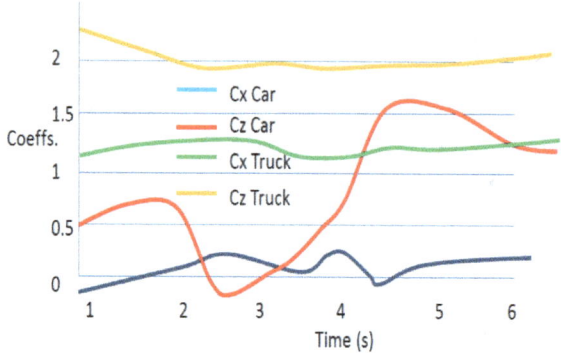

LOW PRESSURE BULB

We have chosen to include this in a separate section so that you are aware of its importance.

When asked if increasing the frontal area will increase or not drag, the majority would answer that yes. In some way it is true, but keep in mind what's behind...

Take for example, a plane like the Jumbo 747. In the front upper part we can see a bulb like geometry:

This bulb geometry is used to include "extra" passengers but the fundamental reason lies in reducing aerodynamic drag and therefore in reducing fuel consumption. So that the plane has more autonomy, which is essential for airline companies.

The most important premise in terms of drag reduction to be taken into account in any racing car design is:

- Fill "physically", areas of low pressure.

Let's take a look at the following example: suppose that we work in 2D with a rectangular body, where the air goes from left to right; the goal is to modify the walls' geometry to reduce aerodynamic drag;

To find which the pressure distribution on the walls is and see where the low and high pressure areas are located, we proceed to perform a CFD simulation, obtaining the following pressure map:

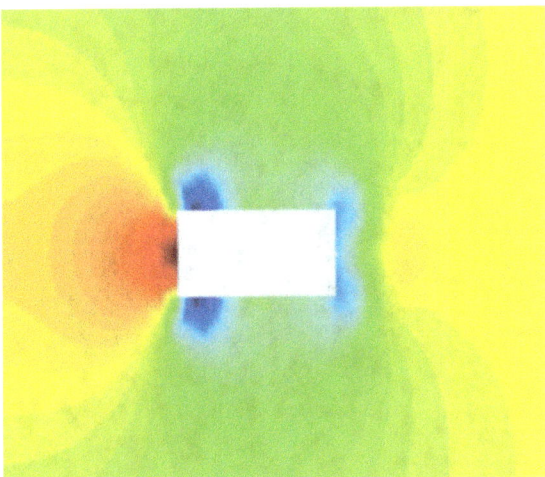

The red color indicates high pressure (air collision in this case), and dark-blue color indicates low pressure.

We can clearly see a well-defined low pressure area in the front edges.

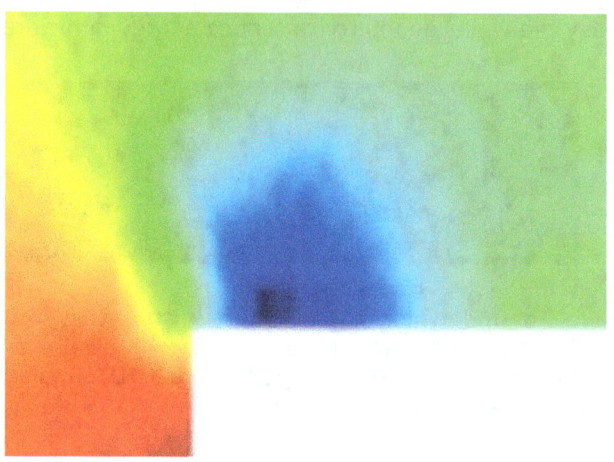

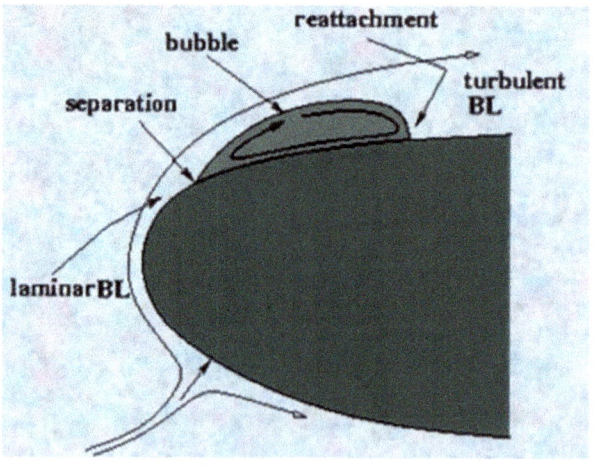

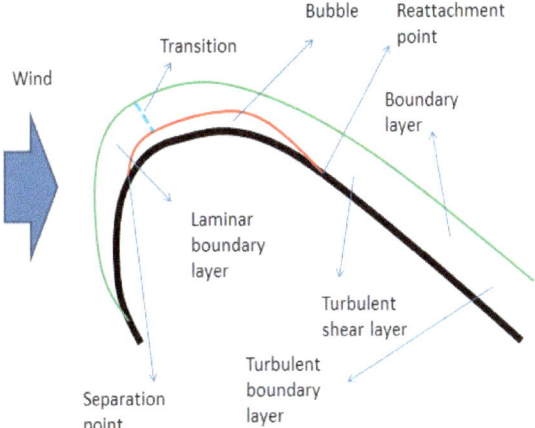

The goal of this design method is to fill this area conveniently; i.e.: make a solid wall. This would increase the frontal area but would reduce drag:

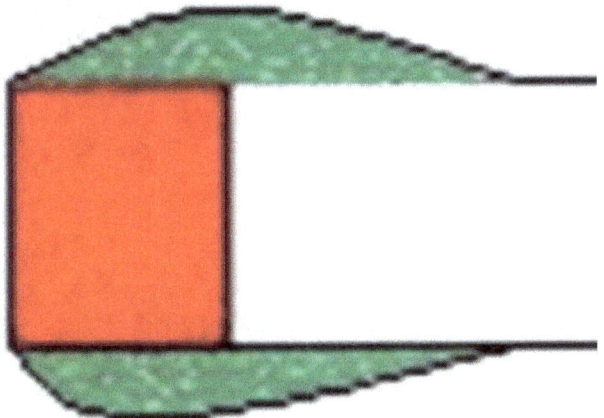

To design these bulb shapes, we have calculated the pressure on the wall, obtaining the following graph (note that the mesh is very poor, thus having few calculation points):

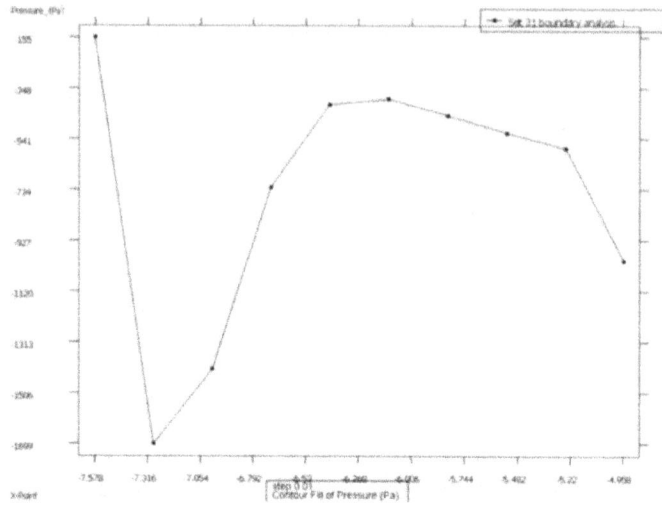

The design process starts right here; we can see how the curve is. Let's do this: let's create a symmetrical mirror-image of the graph, from top to bottom:

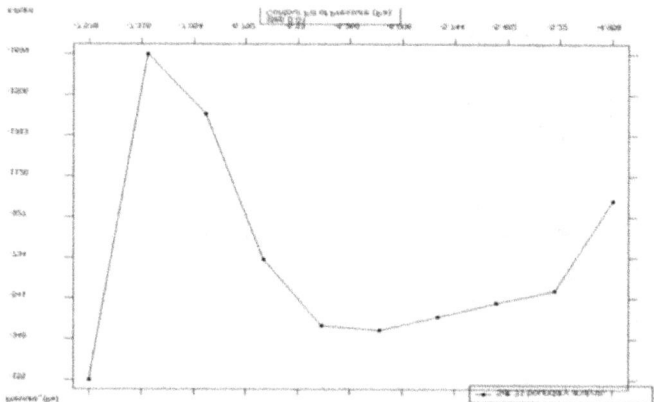

The resulting graph approximates the shape we give to the wall to reduce drag. The following image is exaggerated, but indicates what we need to do:

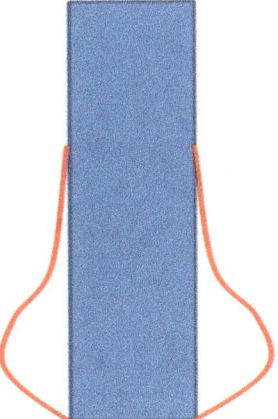

We must now find the bulb height and length; This is an iterative method:

We first build a CAD design and transfer it to a CFD model; with the results obtained and from the pressure map, we vary the geometry and generate another CAD for testing; In our experience, about 5 iterations is sufficient to have the correct and optimal result.

The pressure map will be different in each simulation as the geometry will have changed, but we must build on this new map to optimize and modify the geometry.

Note:

We can also appreciated an area of low pressure in the rear; this means that we should also change the rear's geometry to reduce drag:

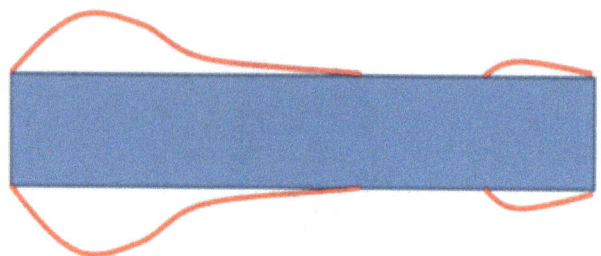

If we extend the bulb concept to the entire geometry we end up with a simetric wing profile:; This is a very important thought:

It is important to know that these bulbs, their size, position and number or amount, <u>depend on the size of the body to optimize.</u>

In a racing car, how can we know the area or areas where there's low pressure and where there is a loss of power?

We can do this in several ways:

1) With pressure sensitive paints; these paintings mark the zones of low or high pressures; we can apply them on track tests and in a wind tunnel;

2) Using CFD: low pressure areas or with large gradients, almost always coincide with areas where the flow separates from the surface. To visualize them we can use various methods: one is to use CFD techniques, using what is called "Oil Flow"; It's like covering the car in oil and see which are the flow lines "on" the surfaces.

3) Substituting the surface to study for canvas or a similar malleable material. Using a camera mounted on the car we can see the movement of the canvas, indicating areas of low or high pressure, even turbulence!

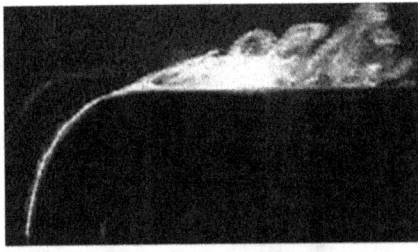

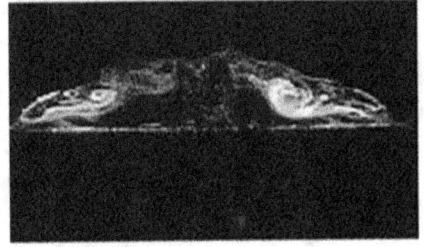

We can see this depression bubbles in sidepod of F3:

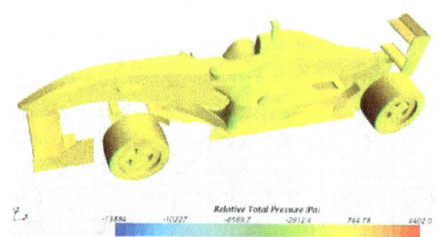

→ Rear ears and front bulbs are typically used to reduce drag.

We can, however, choose another method; instead of filling low pressure areas with bulbs, we can try to mitigate them rounding the front as we can see in the following images. The effects are clear, we can see a reduction in:

- The low pressure in the front.
- The thickness of the turbulent boundary layer on the roof of the truck.
- The low pressure in rear area.

This represents a large reduction in drag.

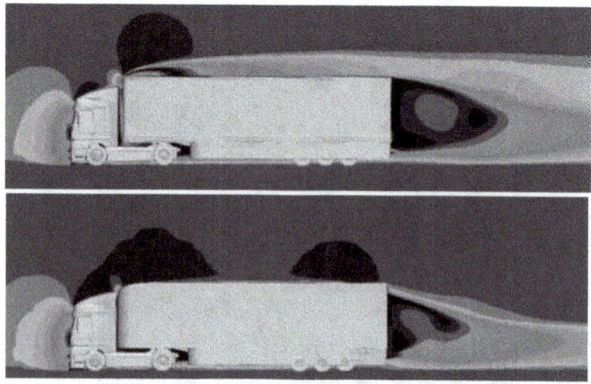

With this system, we fill the depression in the rear, without reducing the payload of the truck, modifying the rear geometry:

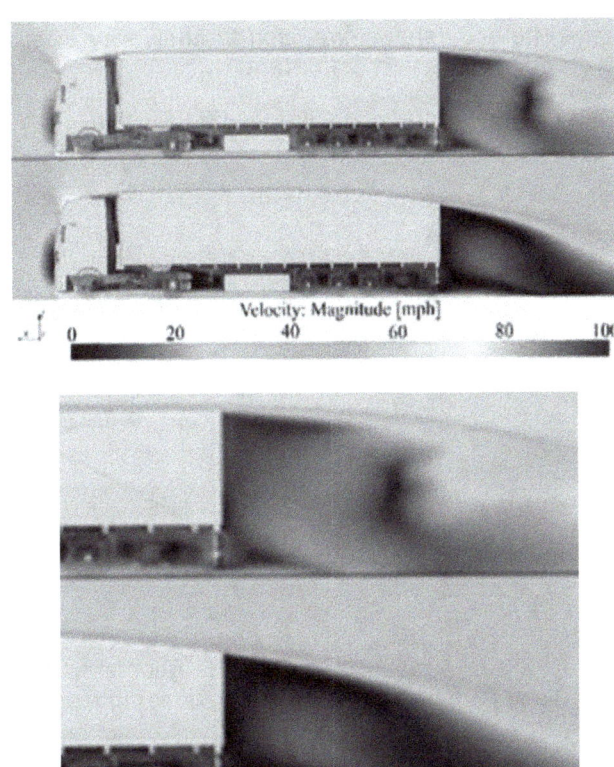

Velocity: Magnitude [mph]

0   20   40   60   80   100

→ Note:

For example, have a truck; hi have 2 zones (high and low pressure):

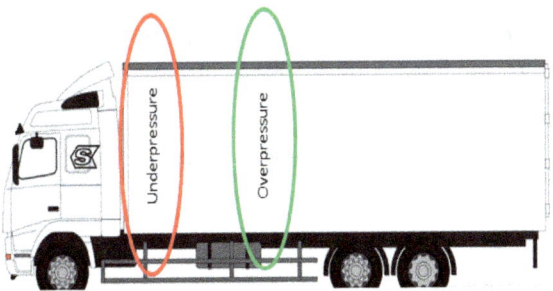

If analyze it, and if we want to optimize and improve the body shape, we need to know why exist the zone green. That is very important.

If we thinking about the body ideal (low drag):

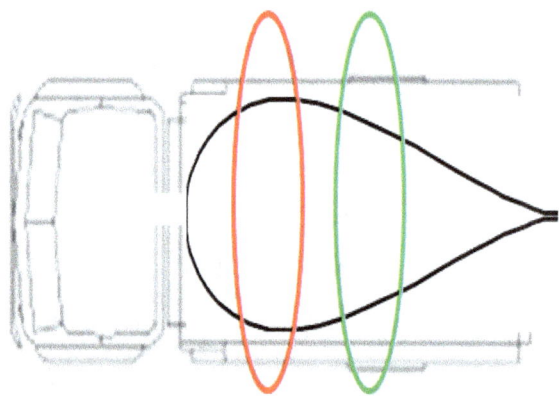

The air wants to adapt to the geometry of less drag. This is the reason why there is a high pressure zone.

JOINTS BETWEEN SURFACES AND SCREWS

Joints between surfaces and screws is something important and we want to cover it in this section, for its uniqueness and impact.

We have probably heard that CFD models, have an error of 5%; where does this value come from? Indeed CFD has an error but not an error attributable to the CFD itself, the mistake comes from the person who programs the simulation, and especially from the person who draws the car.

When we go and simulate a car using CFD, we need to have the car in a CAD model; the more accurate the geometry of the car is, the more memory and time will be required to perform the study; for accuracy, we are not referring to accuracy of the computational simulation but accuracy of detail and reflection in CAD of the reality: we sometimes forget to include in a CAD model joints between surfaces or screw heads. If we do not include them, we are simulating "something" that is not real, because the car in CAD will not match the real car; hence the error: CFD simulates the CAD that is being imported.

In both problems described, it is necessary to know the best options for attaching surfaces and leave the heads of the screws visible. In the following table we can see the resistances of different types of connections between surfaces, in one direction and in another:

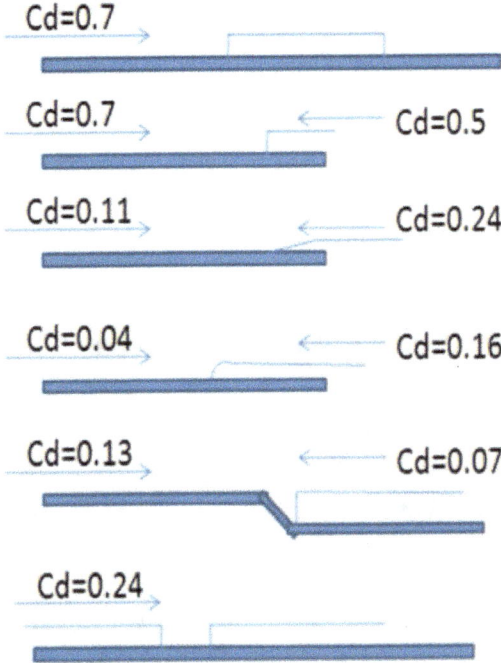

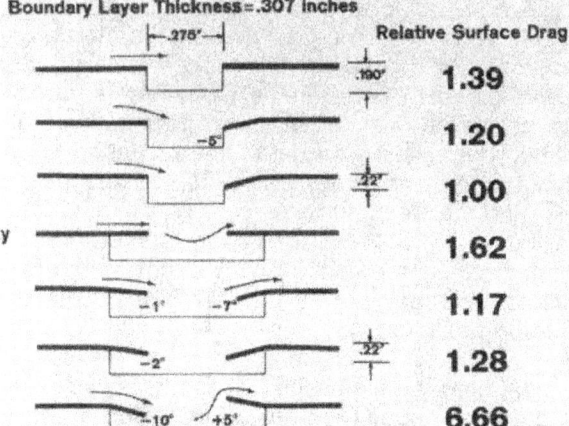

Boundary Layer Thickness=.307 inches

Relative Surface Drag

1.39

1.20

1.00

1.62

1.17

1.28

6.66

The resistance of different types of screw head:

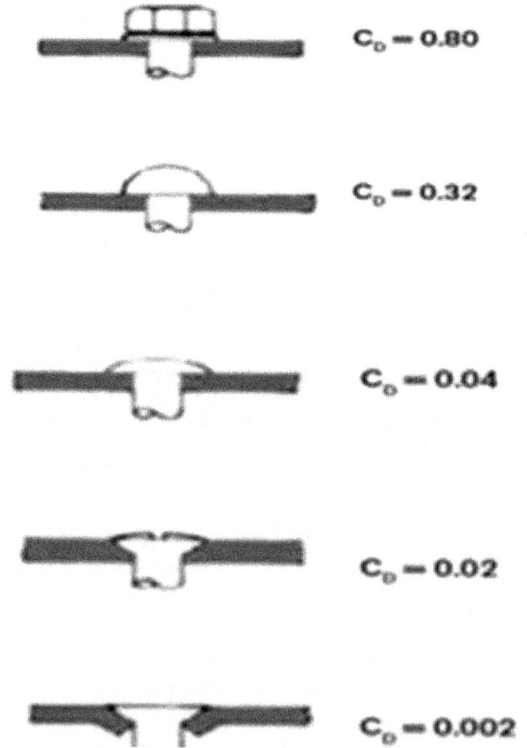

$C_D = 0.80$

$C_D = 0.32$

$C_D = 0.04$

$C_D = 0.02$

$C_D = 0.002$

As a curiosity, look at the joints turbulators sidepods between Red Bull and McLaren: the difference in smoothness, is very clear:

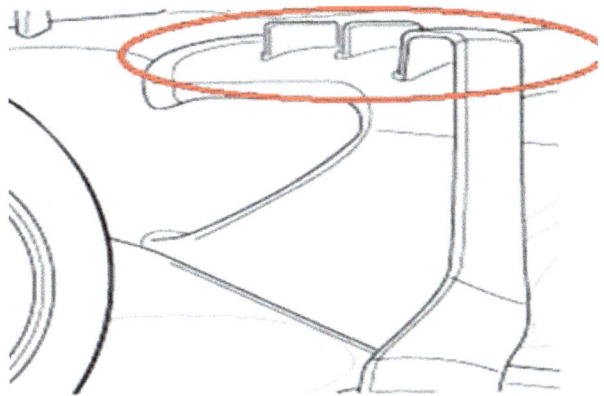

The joints, in the case of Red Bull, are meticulous, smooth and perfect; not so in the case of McLaren.

DRAG REDUCTION OF WHEELS

A clear practical example on how to reduce the global drag in a car is to focus on its wheels; the wheels on a open wheeled formula car represent approximately 40% or more of the total resistance of the car; This means that any "small" action on the wheel's flow, implicates a very significant reduction of total drag.

Years ago, when the Formula 1 regulations allowed many tests, some cars appeared with 6 wheels to increase the amount of rubber in contact with the track, reducing drastically drag:

We have 2 options when it comes to reducing wheel drag:

1) Use the front wing to deflect the airflow around the wheels.

Use the front wing, to place a channel that collects air from the front and place it on the rear of the wheels filling the "hole" of depression that is formed:

This procedure is the same that Ferrari used in 2013 season.
Another system is based from nut wheel; the flow air is the refrigeration brakes:

Another important system to reduce the drag of the wheels is called strakes: These are placed in the bottom of the front flap (F3 Dallara 308, for example); see later other consequences or functions:

They have a height between 2 and 3 cm and a length of 28 cm.
—> Important:

If we are able to reduce the drag in wheels, we are able to:

1) Improve the maximum velocity.
2) Improve the grip between the tire and the track, because we are able to reduce the wheel-generated lift; as a consequence we improve the downforce; improving tire grip.

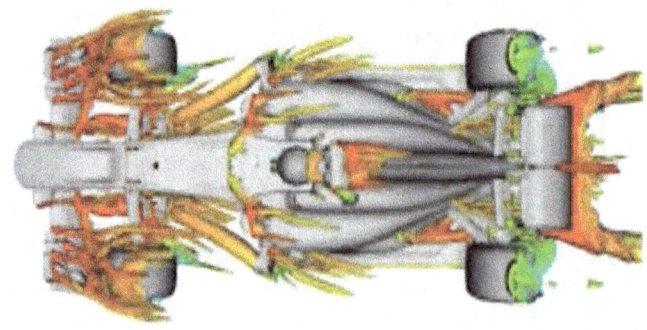

2) Wheel fairings remove air from the high pressure region within the fairing.

Inside the fairing, a collision of several airflows takes place:

  a) From the front.
  b) From the brake and cooling.
  c) From the inside.

All these flows inside coincide, producing high pressure and turbulence, this increases drag dramatically; Solution: remove air from this area.

This extraction is performed by "louvres"; There are many types depending on the characteristics of the car or even the rules of the racing series, but their basic goal is to remove the air and relieve pressure; obviously another major area of work is to find an appropriate way to channel the exhaust air...

These extractors can be placed in different areas, such as behind and above the wheels:

"Louvres", may be designed in a simpler way, but the efficiency is the same; depends on the overall design of the car, as always:

They may also be so large that they fully open the top of the fairing:

Sometimes, these louvres may be confused with openings for cooling:

In short: we divert the air trying to generate the greatest amount of downforce; for this, we can divert the flow up or into a part of the car where it is needed:

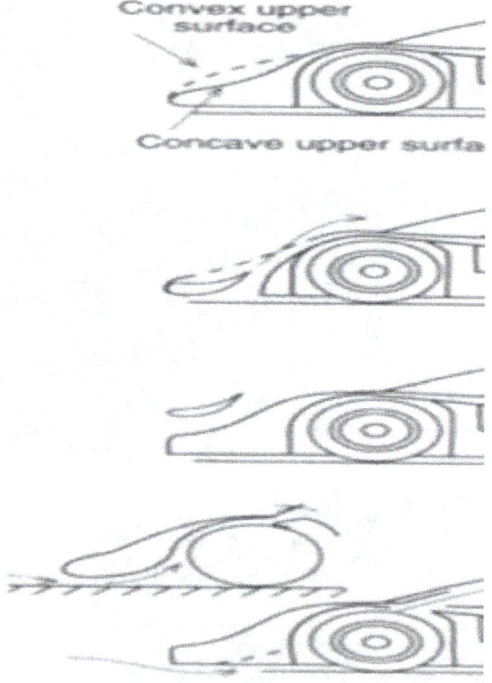

We know too already, that the wheels have a lot drag; more or less the 40% or more, full darg car iiii; for that, is very important reduce this drag, from devices; as examples:

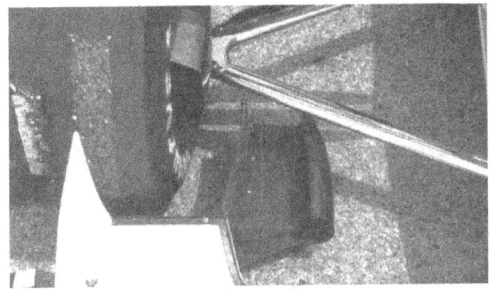

In general too:

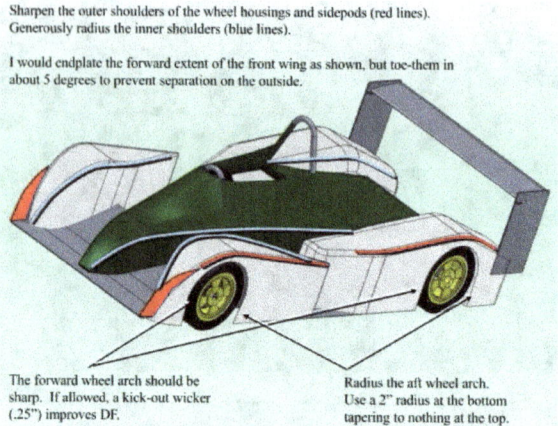

Sharpen the outer shoulders of the wheel housings and sidepods (red lines). Generously radius the inner shoulders (blue lines).

I would endplate the forward extent of the front wing as shown, but toe-them in about 5 degrees to prevent separation on the outside.

The forward wheel arch should be sharp. If allowed, a kick-out wicker (.25") improves DF.

Radius the aft wheel arch. Use a 2" radius at the bottom tapering to nothing at the top.

In the actuality, there are many research in drag reduction in wheels (pneumatics); for example, Yokohama tyres, have a system, which place in tyre some turbulators; that produce in rotation, one reduction in full drag and one flow control around the car:

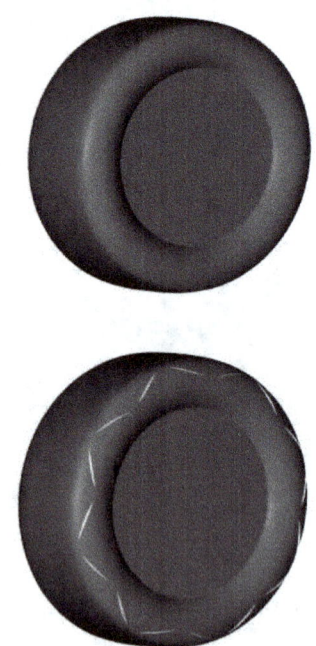

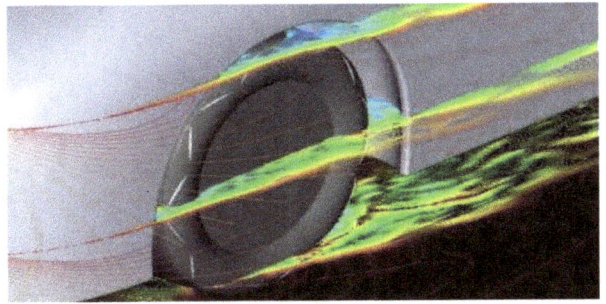

ARCHITECTURE – SPECIAL APPLICATION

The reduction drag in this point is very important:
- Reduction forces.
- Reduction vibrations.

We will see some examples about:
1. Drag reduction:

 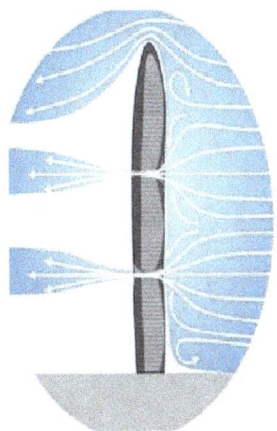

2. Dag reduction and vibrations 25%:

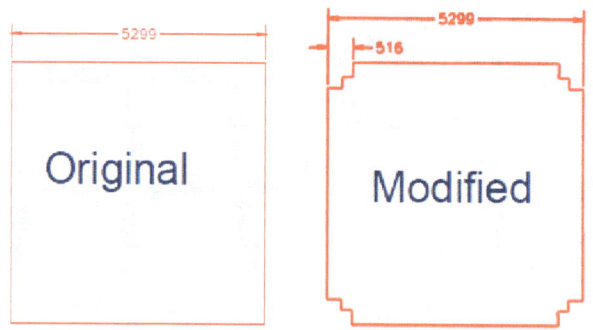

3. Drag reduction and vibrations:
4.

EFFICIENCY L/D

We can generate a lot of downforce but in return we get too much drag, shortly we will be moving towards "efficiency". The value "L / D" is called efficiency or performance, and of course, the higher the better. Today in F1 we reach efficiency values around 3; Chapman, with their innovative designs reached values of 7 and even 8! It is a good practical exercise to get a profile with a high efficiency value. Given a profile, we can represent the variations of its "efficiency" as a function of the Reynolds number:

$$Eficiency = Ef = \frac{L}{D}$$

MOMENTS OF INERTIA AND SPEED OF A CAR

When modelling the behaviour or dynamics of a racing car, we must have completely configured the car to study; This means we must have aerodynamic values, the engine curves, the transmission relations, etc ....

But we also need the moments of inertia of the car even though it is not a strictly aerodynamic issue, but we know how convenient it is to calculate them. The moments of inertia depend on the mass distribution of the car; There is a parameter called Dynamic Index (we'll see), which marks the suitability of the car in yaw relative to its design and its dynamics.

We know the definitions of yaw, roll and pitch:

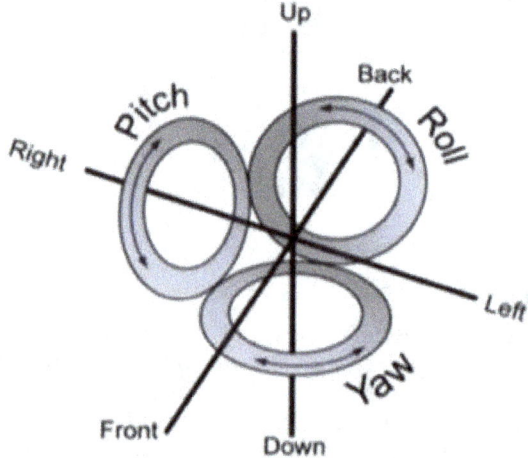

Although we could have completely drawn and designed the car on our computer and even though CAD programs may calculate moments of inertia there is nothing better than reality. In the case of not having the car at a 1:1 scale, we will calculate it with a computer; otherwise, we can choose the following options:

a) Pay and go to a specialised centre where it can be calculated.
b) Use our ingenuity (as engineers) to calculate them.

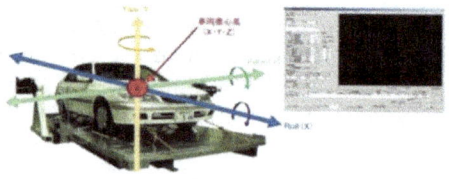

What is this process form platform:

Method of the oscillating platform

Determination of pitch and roll moments of inertia can be carried out measuring the frequency of the oscillation of the car + platform group. (Figure 8- Method of oscillating platform (Source: Vehicle Dynamics documentation)Figure 8). The device can be designed in order to hold the car on the platform either longitudinally and transversally. This way, both moments of inertia can be determined with the same platform. The method is based on the differential equation governing the oscillating movement of the platform. Without exciting and friction forces, potential energy plus kinetic energy must remain constant. They are defined as:

$$E_c = \frac{1}{2} I \cdot \dot{\varphi}^2$$

$$E_p = m \cdot g \cdot z (1 - \cos\varphi)$$

System's total energy will be obtained adding up both expressions, and its temporal derivative must equal 0.

$$(I \cdot \ddot{\varphi} + m \cdot g \cdot z \cdot \varphi)\dot{\varphi} = 0$$

The equations governing rolling and pitching oscillations will be, respectively:

$$[I_{XX} + I_p + m_V \cdot z_V{}^2 + m_P \cdot z_P{}^2]\ddot{\phi} + [m_V \cdot z_V + m_P \cdot z_P]g \cdot \phi = 0$$

$$[I_{YY} + I_p + m_V \cdot z_V{}^2 + m_P \cdot z_P{}^2]\ddot{\varphi} + [m_V \cdot z_V + m_P \cdot z_P]g \cdot \varphi = 0$$

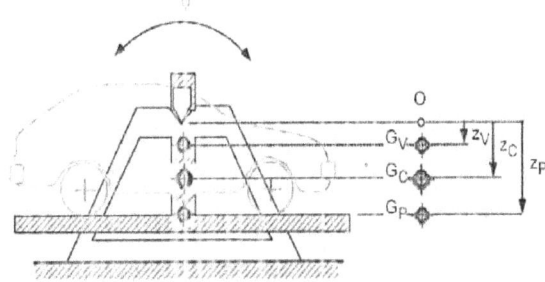

Figure 8- Method of oscillating platform (Source: Vehicle Dynamics documentation)

Aligning car and platform CG on the same vertical, oscillation period measurement allows to calculate the rolling or pitching moment of inertia of the vehicle. Taking mass moments relative to the oscillating axis, the following equations are obtained:

$$z_C \cdot m_C = z_P \cdot m_P + z_V \cdot m_V$$

$$m_c = m_p + m_v$$

$$\tau_p = 2\pi \sqrt{\frac{I_{P,0}}{m_p \cdot g \cdot z_p}}$$

$$\tau_c = 2\pi \sqrt{\frac{I_{c,0}}{m_c \cdot g \cdot z_c}}$$

Applying the superposition principle and Steiner's theorem, the moment of inertia of the set relative to the point "0" of oscillation is:

$$I_{c,0} = I_{V,0} + I_{P,0} = (I_v + m_v \cdot z_v{}^2) + (I_p + m_p \cdot z_p{}^2)$$

Finally, clearing the moment of inertia of the vehicle I_V:

$$I_v = \left[\left(\frac{\tau_c}{2\pi}\right)^2 - \left(\frac{\tau_p}{2\pi}\right)^2\right] m_p \cdot g \cdot z_p + \left[\left(\frac{\tau_c}{2\pi}\right)^2 - \frac{z_v}{g}\right] m_v \cdot g \cdot z_v$$

Being z the distance between rotation point 0 to the CG of the mass considered, and τ_c and τ_p the oscillating periods of the set and the platform respectively.

In the second case (the first is easy), do the following:

Hang the car to the roof: knowing the length of the cable where it hangs and knowing the moment of inertia of the platform on which it rests, we can balance the car lengthwise, to calculate the pitch:

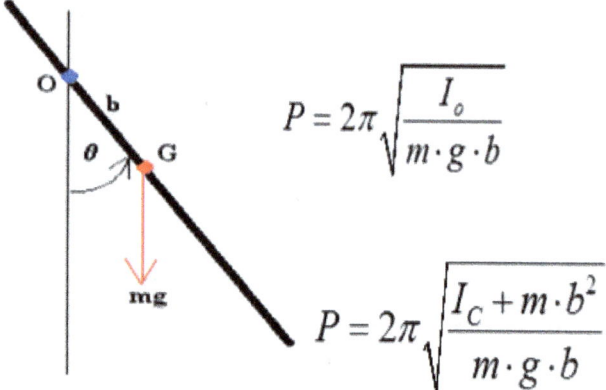

$$P = 2\pi \sqrt{\frac{I_0}{m \cdot g \cdot b}}$$

$$P = 2\pi \sqrt{\frac{I_c + m \cdot b^2}{m \cdot g \cdot b}}$$

Being "P" the period of oscillation or rolling in seconds, and "I" the moment we are looking for.

Similarly, we calculate the roll moment, pushing "laterally" the car:

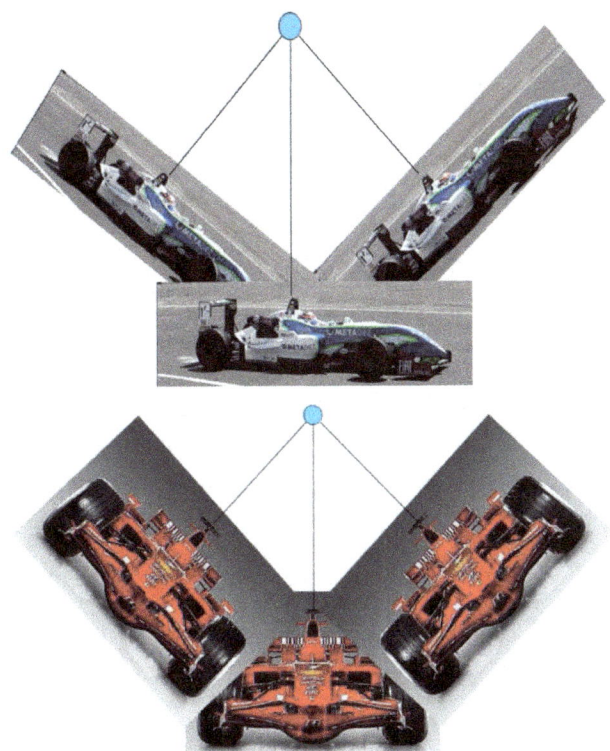

To calculate the yaw moment it is not that easy; so we must hang the car from a torsion bar and make the car turn; We can calculate and measure the period of oscillation; "K" is the constant of the torsion bar:

$$T = 2\pi \sqrt{\frac{I}{K}}$$

If we think a little more, we could design a system that could measure all moments; hang the car on it through cardan placing a sensor on each axis (gyro sensor) and the cardan would be hanged on a torsion bar:

Finally, we can use another method (only valid for roll and pitch) which is perhaps the cheapest of all, because we use the acquisition system and the sensors installed in the car without spending any extra money. Note very important: form roll and pitch moment, is possible to know automatically, the yaw moment iiii.

We can sit on the tip of the nose of our car and get up abruptly. The oscillating data acquired and properly treated will give us the pitch; likewise seated on the side we would be able to calculate the roll moment.

There are other experimental methods to calculate these moments:

1. Method
Pitching moment = 0.99W - 1.149
Rolling moment = 0.18W - 150
Yaw moment = 1.03W - 1.206
"W" is the car's weight in pounds.

2. Method
Now the moment Kg / m^2 and the "W" in kilograms:
Pitch = 2.56 W - 1103
Pitch masses suspended = 2.18 W - 1006
Yaw = 2.86 W - 1315
Yaw masses suspended = 2.42 W - 1198
Rolling suspended mass = 0.28W - 71.6
Rolling unsprung mass = 0.37W - 86.4

The "DYNAMIC INDEX" that we mentioned previously, determines the dynamic understeer or oversteer of the car; it will be analysed and quantified later.

Consider now an important case: suppose we have a wing that generates some lift; we must know how to calculate the aerodynamic interactions between moments (I) which produces the profile, the originating loads (F), the centre of pressure, the point of attachment to the chassis and its distance to the profile (d) and the torque (T):

We want to calculate F_{zr} and F_{zf} from F_x and F_z

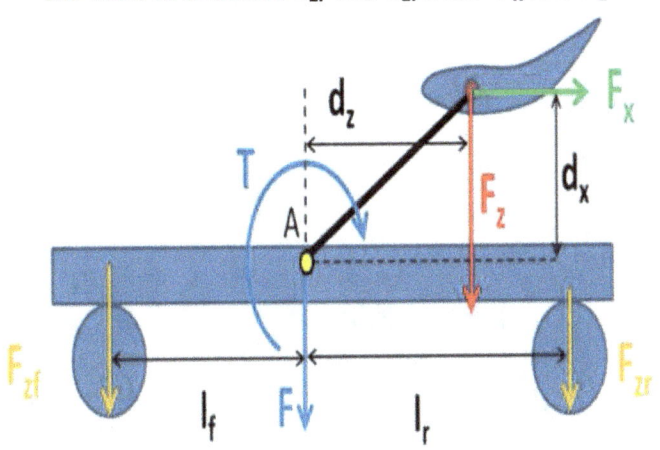

F_x	Drag
F_z	Downforce
F_{zf}	Front load
F_{zr}	Rear load
T	Torque in A
F	Force in A

We assume equilibrium conditions

Force:
$$F = F_z = F_{zf} + F_{zr}$$

Moments
$$T = F_z \cdot d_z + F_x \cdot d_x = F_{zr} \cdot l_r - F_{zf} \cdot l_f$$

We have two equations with two unknowns (Fzr and Fzf)

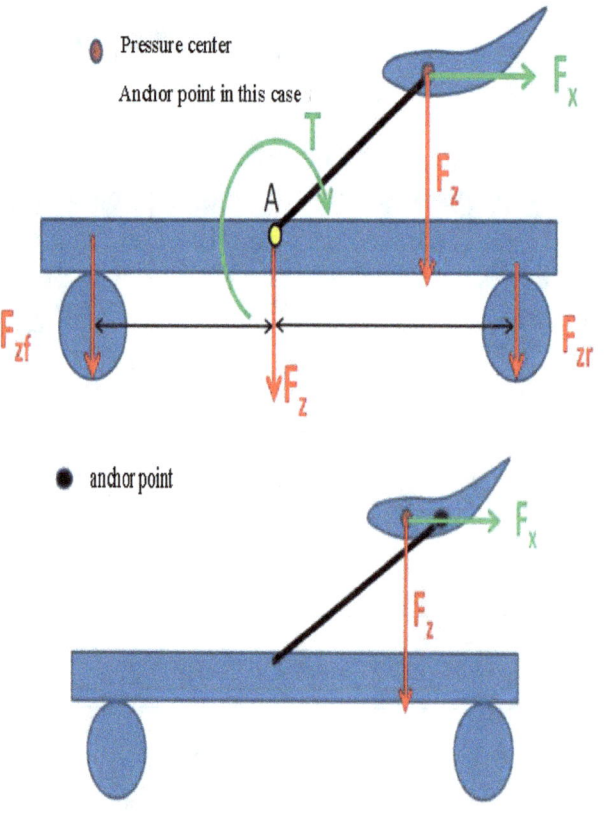

Assume that the anchor point does not allow the rotation of the wing.
In this case the forces in the pressure centre are transmitted unchanged to the anchor point. We would do as in the first case.

SPEED LIMITS IN TESTS TO EXTRAPOLATE

Sometimes we have speed limits in wind tunnels or even in straight line test; We can perform a test at a speed and extrapolate these values at other speeds.

The minimum speed at which we perform on tests, should be one for which data can be extrapolated; logical. But it goes beyond this obvious and depends on the design that the vehicle has.

We can have a car in which forces appear after a certain speed; if this happens, and not surprisingly, we perform several tests at different speeds to find the transition point.

Therefore, we must ensure that we can extrapolate; a suitable speed for it, and knowing the limitations or problems that may appear, this could be 200 km/h.

EXPERIMENTAL METHODS FOR CALCULATION LIFT AND DRAG

⇒ Note:

The shape that "doesn't alter" the air along its trajectory, will generate no DRAG.

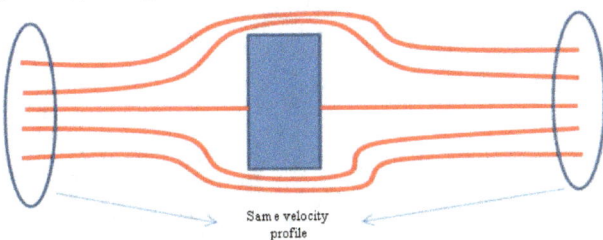

Same velocity profile

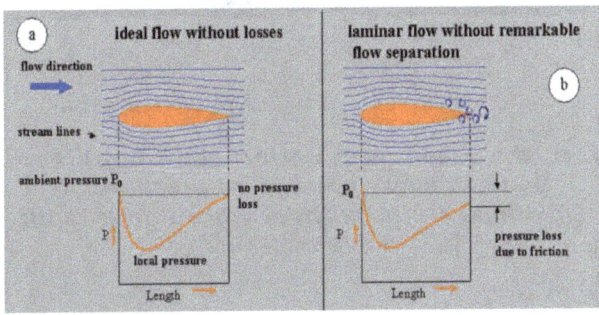

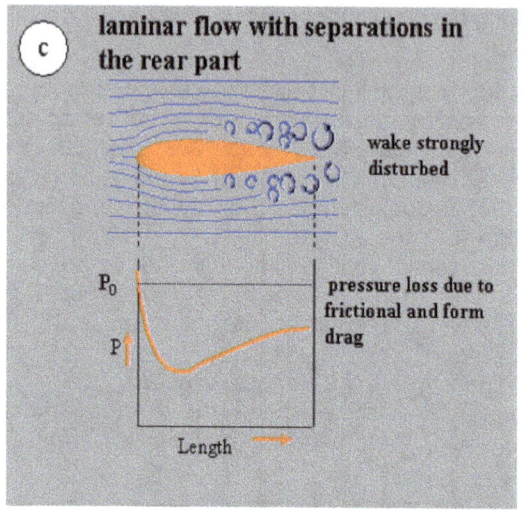

c laminar flow with separations in the rear part

wake strongly disturbed

P_0

P

pressure loss due to frictional and form drag

Length

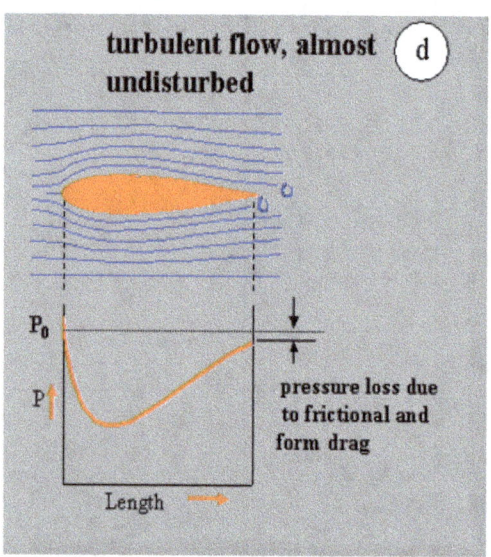

d turbulent flow, almost undisturbed

P_0

P

pressure loss due to frictional and form drag

Length

In fact, the shape most perfect (low drag) is which have in every point, zero pressure....
(it will be static, without vibration, without movement....).

The less energy required to move an object around air the smaller the drag will be. This is a simple principle that describes the dissipation of energy.

If we knew the velocity profile at the front and rear of a car, we could obtain the generated drag. The velocity profile incident to a car can be considered to be laminar. However the velocity profile on the rear of the car is somewhat more complicated and we have to calculate it with CFD simulations or wind tunnel testing.

The following reasoning is perfectly applicable to 3D without problem.

1

Taking 2 velocity profiles, we could design several tests to experimentally determine a function that calculates the drag of any object.

As we said above, were we able to calculate the flow's variations, we would know the drag value; moreover, if we calculate the flow's variation in the vertical axis we get downforce.

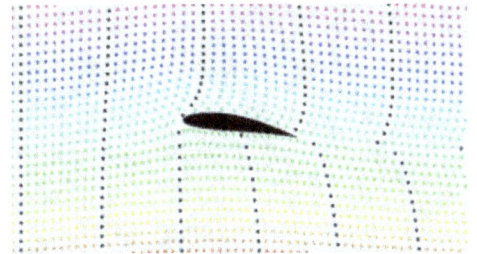

Force equals mass times acceleration; mass is equal to density times volume; We can represent the acceleration as the derivative of the velocity with respect to time;

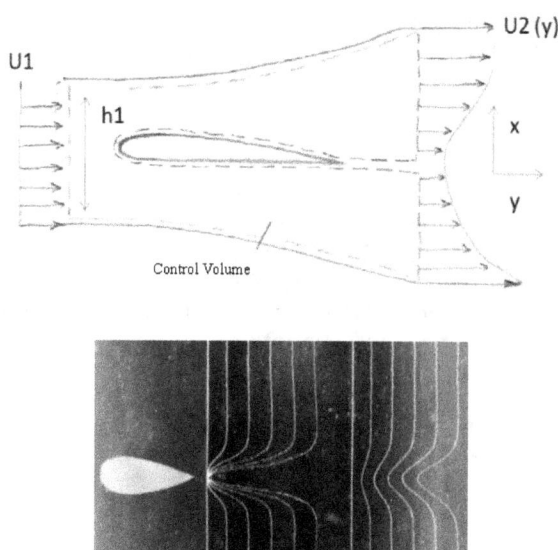

In addition to other logical considerations, we assume that pressure "P" equals "P8"; "B" is the profile's wingspan; "FH" is drag and "FV" downforce:
Continuity equation:

$$\dot{m}_1 = \dot{m}_2$$
$$\dot{m}_1 = \rho \cdot u_1 \cdot A_1 = \rho \cdot u_1 \cdot (h_1 \cdot b)$$
$$\dot{m}_2 = \rho \cdot b \int u_2(y)\, dy$$

Momentum conservation:

$$\dot{M}_{x,out} \prec \dot{M}_{x,in}$$

$$F_H = \dot{M}_{x,out} - \dot{M}_{x,in}$$

$$\dot{M}_{x,in} = \rho u_1^2 \left(bh_1\right)$$

$$\dot{M}_{x,out} = \rho b \int_2 u_2^2 (y)dy$$

$$-F_H = \boxed{D = \rho b \int_2 u_2^2 (y)dy - \rho b h_1 u_1^2}$$

Consider the case where the control volume walls are parallel to the "x" axis :

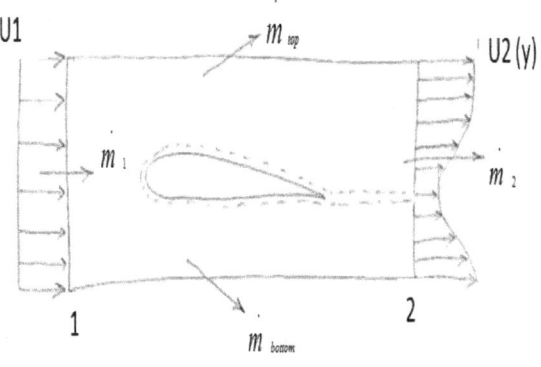

$$A_1 = A_2$$

$$\dot{m}_1 = \dot{m}_{top} + \dot{m}_{bottom} + \dot{m}_2$$

$$\dot{m}_1 = \rho \cdot u_1 \cdot A_1 = \rho \cdot u_1 \cdot (h_1 \cdot b)$$

$$\dot{m}_2 = \rho \cdot b \int u_2(y)\, dy$$

$$\dot{m}_{top} + \dot{m}_{bottom} = \rho u_1(h_1 b) - \rho b \int_2 u_2(y)\, dy$$

$$\boxed{\dot{m}_{top} = \dot{m}_{bottom} = \frac{1}{2}\left[\rho u_1(h_1 b) - \rho b \int_2 u_2(y)\, dy\right]}$$

Momentum:

$$F_H = \dot{M}_{x,out} - \dot{M}_{x,in} = \dot{M}_{x,out,2} + \dot{M}_{x,out,top+bottom} - \dot{M}_{x,in}$$

$$\dot{M}_{x,in} = \rho u_1^2(bh_1)$$

$$\dot{M}_{x,out} = \rho b \int_2 u_2^2(y)\, dy$$

$$u_{top} = u_{bottom} \simeq u_1$$

$$\dot{M}_{x,out,top+bottom} \simeq 2\dot{m}_{top} u_1$$

$$-F_H = \boxed{D = \rho b \int_2 u_2^2(y)\, dy + 2\dot{m}_{top} u_1 - \rho b h_1 u_1^2}$$

Let's finally look at how downforce and drag are calculated in a wind tunnel:
Continuity:

$$\dot{m}_{in} = \dot{m}_{out}$$

$$\rho u_1 A = \rho \int_2 u_2(y)\,dy = \rho \bar{u}_2 A$$

$$u_1 = \bar{u}_2$$

$$\bar{u}_2 = \frac{1}{h}\int_2 u_2(y)\,dy \quad and \quad \overline{u_2^2} = \frac{1}{h}\int_2 u_2^2(y)\,dy$$

$$\bar{u}_2^2 \neq \overline{u_2^2} \Rightarrow \dot{M}_{x,1} = \rho u_1^2 A \neq \rho \overline{u_2^2} A = \dot{M}_{x,2}$$

Momentum in "x":

$$-F_H = \boxed{D = \rho b \int_2 u_2^2(y)\,dy - \rho b h u_1^2 + (p_2 - p_1)bh}$$

Momentum in "y":

$$F_V + b\left(\int_{Lower\,CS} p_l(x)\,dx - \int_{Upper\,CS} p_u(x)\,dx\right) = \dot{M}_{y,out} - \dot{M}_{y,in} = 0$$

$$-F_V = \boxed{L = b\left(\int_{Lower\,CS} p_l(x)\,dx - \int_{Upper\,CS} p_u(x)\,dx\right)}$$

→ Sample (from Alejandro Murillo Juliá, end project study):

A wake rake probe is a device which is used for obtaining the drag by studying the flow conditions downstream the wind tunnel test model

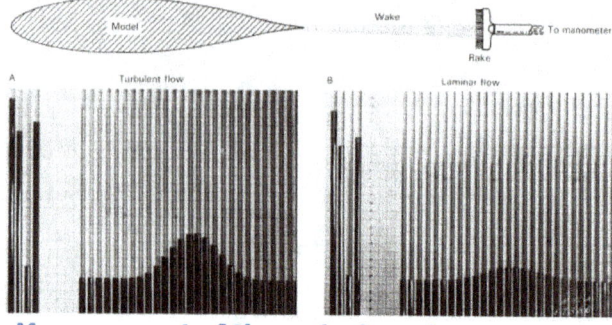

Measurement of the wake by using a pressure rake.

A wake rake probe has a number of static probes and dynamic pressure probes, which allow to obtain the distribution of the total pressure downstream. Knowing the distribution of the total pressure and the value of the static pressure in the wake, it is possible to obtain the distribution of the dynamic pressure and, then, the distribution of the velocity field downstream the model. This allow to obtain the drag coefficient by studying the difference in the amount of movement of the flow.

A possible configuration for a rake is shown next. In this case, the rake has some static and dynamic probes with their axis placed in the same direction as the free stream. The probes must be pointing to the inlet of the wind tunnel and the dynamic ones will provide the dynamic pressure measurements. Applying the equation of Bernoulli, the total pressure will be obtained.

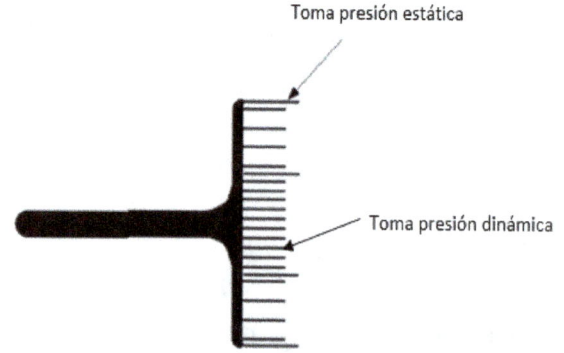

Pressure rake with static and dynamic pressure probes.

For obtaining the pressure measurements, the probes are connected to a scanivalve that is linked to a computer that receives the processed data, provided by the pressure transducer. Next, a general scheme of the configuration of the instrumentation of a rake is attached.

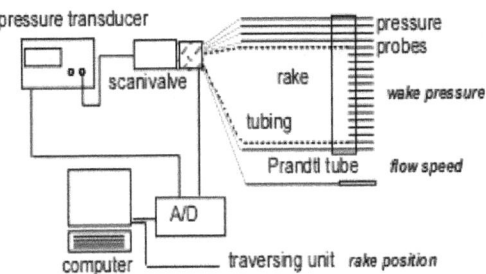

General scheme of the configuration of the instrumentation of a rake.

Once the pressure gradient downstream the model has been obtained, the width and the state of the boundary layer can be known, also the distribution of velocity downstream the model and, then, the total drag. Due to the accuracy and the ease of use of these devices, the rakes are used in multiple applications, such as:

- Study of the boundary layer in wind tunnel tests of civil aerodynamics.
- Obtain the drag of different profiles, tested in wind tunnel.
- Test the wake generated downstream different models.
- Obtain the drag generated by the geometry of different wings and the wake of the wheels in racing vehicles.
- Analyze the wake generated by different propellers.
- Study the effect of ice in the wings of aircrafts.
- Flight tests.
- Study the drag of different types of birds.

Wheels' wake rake probes set up in a Formula 1 car.

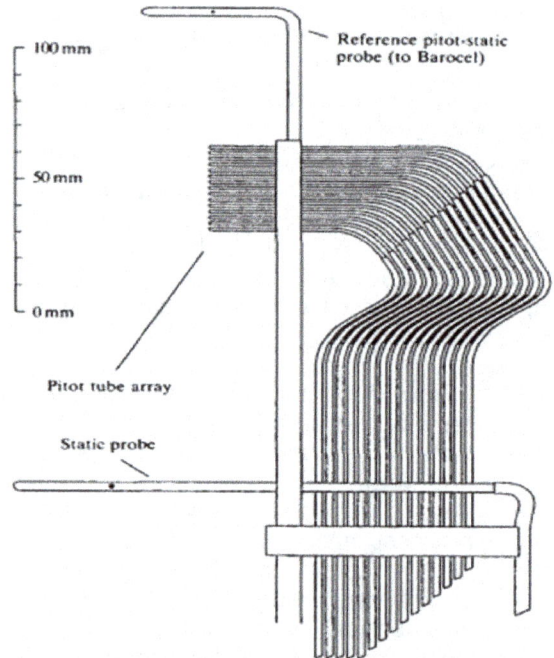

Reference pitot-static probe (to Barocel)

100 mm

50 mm

0 mm

Pitot tube array

Static probe

Rake designed for studying the drag generated by a hawk.

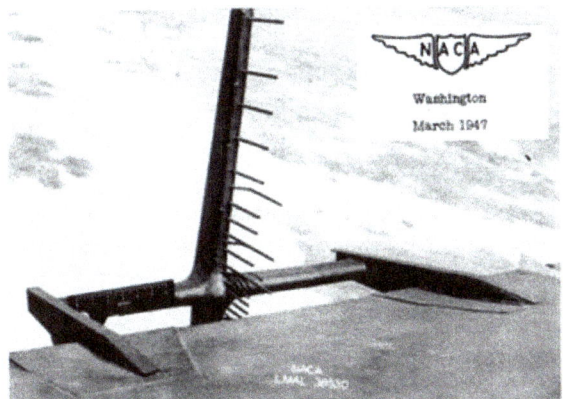

Detailed view of a rake used in flight tests.

Theoretical basis for obtaining the drag coefficient

For obtaining the drag by analyzing the pressure measurements of a wake rake, a reference system, fixed to the wind tunnel, is defined. This reference system can be observed in the next figure:

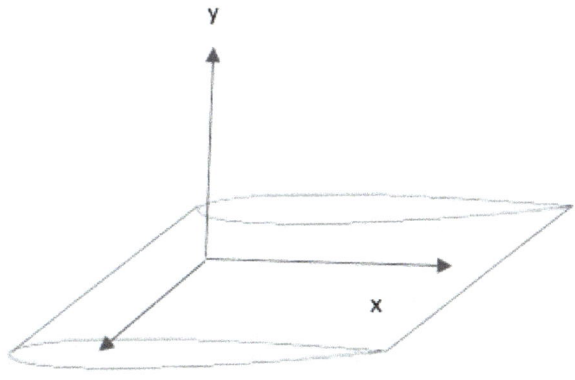

Reference system, fixed to the wind.

The velocity of the stream, in this reference system is defined as the following expression:

$$\mathbf{v} = [U_\infty, 0,0] + [u, v, w]$$

Where the component associated to the free stream can be identified as U_∞ and the perturbations due to the geometry of the model are defined as "u", "v" and "w". As this project is working on a two-dimensional test in which the three-dimensional effects are not considered, the component in "z" can be neglected.

In the following figure, a scheme of the velocity field is shown with its different characteristic sections (at the inlet, after the trailing edge and downstream):

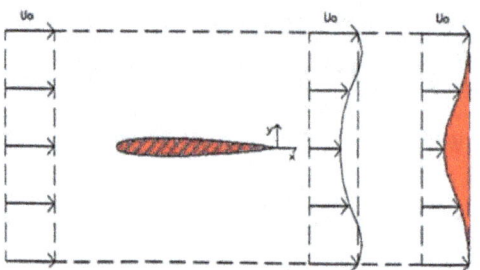

Scheme of the velocity field with its different characteristic sections.

The model generates the boundary layer, whose thickness grows as the flow advance through the geometry of the model until the flow separation takes place, resulting in a turbulent wide wake that modifies the velocity field downstream the model.

From now on, the next hypothesis will be considered:

- Two-dimensional movement $\frac{\partial}{\partial z}=0$
- Stationary motion $\frac{\partial}{\partial t}=0$
- Incompressible movement $\rho = cte$
- $Re \gg 1$
- Friction shear stress can be neglected

Next, two views of the volume of control chosen for studying the test by applying the mechanical of fluids is attached:

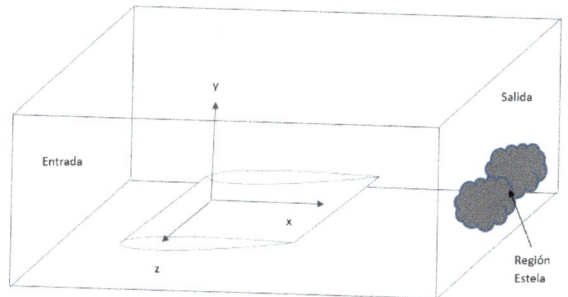

Three-dimensional view of the volume of control chosen.

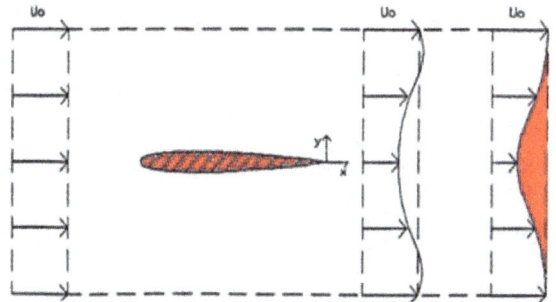

Two-dimensional view of the volume of control chosen.

According to the conservation of mass law, the flow at the inlet must be the same as the flow that exits by the outlet and the upper and lower surface. Then, the difference between the flow at the inlet and outlet is the flow that exit from the volume of control by the upper and lower surface.

Applying the equation of amount of movement to this volume of control, the equation that defines the drag can be obtained. Considering that it is a two-dimensional problem, the expression of the drag will provide a force/length:

$$(S = cl_z = c \cdot 1 = c)$$

$$\frac{d}{dt} \int_{V_c} \rho \mathbf{v} \, dV_c + \int_{A_c} \rho \mathbf{v}(\mathbf{v} \cdot \mathbf{n}) \, dA_c$$

$$= -\int_{V_c} p \cdot \mathbf{n} \, dV_c + \int_{A_c} \tau \cdot \mathbf{n} \, dA_c + f_{grav} + f$$

$$D = \int_e \rho U_\infty^2 \, dy - \int_s \rho U_s^2 \, dy - \int_{sup,inf} \rho U_\infty (U_\infty - U_s) dy$$

Now, the expression of the drag is going to be converted into no-dimensional, by using the dynamic pressure of the free stream and the surface of reference $(S = cl_z = c \cdot 1 = c)$.

$$D = \frac{1}{2} \rho U_\infty^2 S C_D = q_\infty S C_D$$

$$C_D = \frac{D}{q_\infty S}$$

Taking into account that the inlet has the conditions of the free stream and the outlet is the section where the wake rake probe is placed (s=r):

$$D = \rho \int_{l_{rake}} U_s(U_\infty - U_s) \, dy$$

Knowing the expression of the amount of movement boundary layer of the mechanical of fluids, $\delta_{2\infty}$. And extending the range of integration until ∞:

$$\delta_{2\infty} = \int_{-\infty}^{\infty} \frac{U_s}{U_\infty}\left(1 - \frac{U_s}{U_\infty}\right)$$

$$D = \rho U_\infty^2 \delta_{2\infty}$$

$$C_d = 2\frac{\delta_{2\infty}}{c}$$

It must be considered that the wake rake must be placed far enough from the model, allowing the static pressure to reach the same value as in the free stream and, for improving the accuracy, several static pressure probes are set up in different regions of the rake

Geometrical parameters and design.

The following description is based on a research developed by Lloyd N. Krause to determine the effect of pressure rake design parameters on static pressure measurement for the NACA (National Advisory Committee for Aeronautics). The investigation covered a Mach number range of 0,3 to 0,95.

The accurate measurement of static pressure by the-conventional tube depends upon the configuration chosen; that is, the proper location of the static orifices in relation to the nose and the support of the tube. These investigations were limited, however, to single tubes. Pressure rakes require considerable change in configuration from the basic tube design, which results in additional error in the static-pressure portion of the rake. The extent of this error depends mainly upon the five design parameters shown in the next figure. These parameters were:

1. The distance from the static orifices to the leading edge of the static-pressure tube.
2. The distance from the static orifices to the support.
3. The distance between adjacent and static-pressure tubes.
4. The distance from the static orifices to the leading edge of the adjacent tubes.
5. The ratio of support diameter to jet diameter.

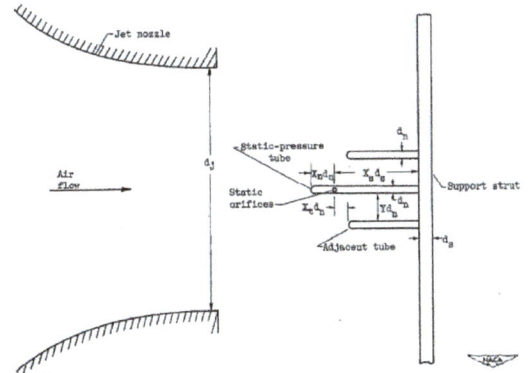

Scheme of the main geometrical parameters in a pressure rake.

The choice of an optimum configuration which would satisfy any test condition desired is impractical because of the restrictions that many installations impose. However, certain recommendations can be stated as a guide to the choice of a rake configuration.

A value for the distance from the static orifices to the leading edge of the static-pressure tube greater than 3 diameters is recommended because the change in pressure coefficient for values greater than 3 diameters is small.

The choice of a value for the distance from the static orifices to support depends primarily upon tube strength considerations. It is desirable that the distance from the static orifices to support be at least greater than 6 support diameters. For values less than 6 diameters, the slope of the pressure coefficient Mach number curve becomes steep at the high Mach numbers and an accurate-estimate of the pressure coefficient becomes difficult. A value of 10 diameters or greater is recommended.

It is desirable to minimize the ratio of support diameter to jet diameter in order to obtain a pressure coefficient close to the infinite stream value.

In considering the proximity effects of adjacent tubes near the static pressure tubes, it is recommended that the tubes be 5 diameters or more apart. For cases where the tubes are less than 5 diameters apart, the leading edge of the adjacent tubes should be in line with the static orifices to minimize the proximity effects on pressure coefficient.

2
DOWNFORCE CALCULATION PROCESS: MESURES, SENSORS AND SYSTEM DATA ACQUISITION; FIRST METHOD:

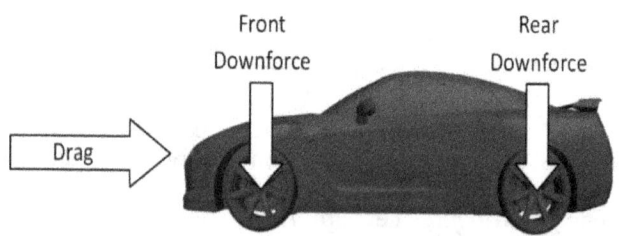

DOWNFORCE CALCULATION PROCESS: FIRST METHOD:
From pressure distribution around wing:

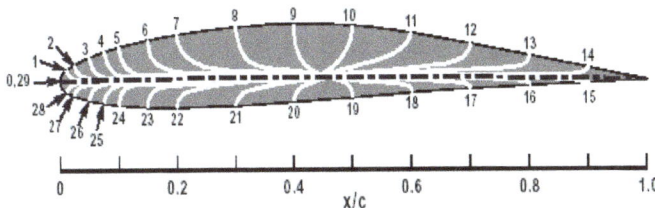

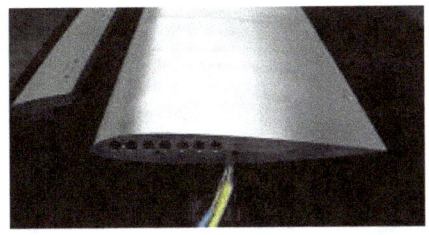

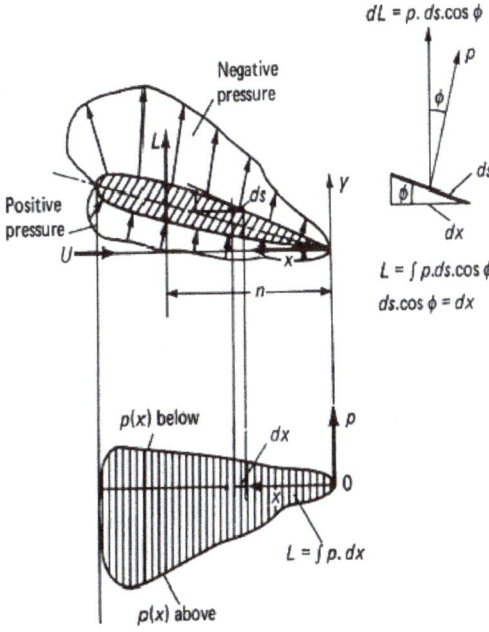

- The lift force L is determined by integration of the measured pressure distribution over the airfoil's surface.
- It is expressed in a dimensionless form by the pressure coefficient C_p where, p_i = surface pressure measured, = P pressure in the free-stream
- The lift force is also measured using the load cell and data acquisition system directly.

U_∞ = free-stream velocity, r = air density (temperature),

$p_{stagnation}$ = stagnation pressure measured at the tip of the pitot tube, L = Lift force, b = airfoil span, c = airfoil chord

$$C_p = \frac{p_i - p_\infty}{\frac{1}{2}\rho U_\infty^2} \qquad U_\infty = \sqrt{\frac{2\left(p_{stagnation} - p_\infty\right)}{\rho}}$$

$$C_L = \frac{2L}{\rho U_\infty^2 bc}$$

$$L = \int_s \left(p - p_\infty\right)\sin(\theta)ds$$

$$C_L = \frac{\int_s \left(p_\infty - p\right)\sin(\theta)ds}{\frac{1}{2}\rho U_\infty^2 c}$$

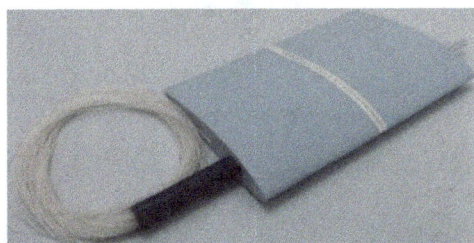

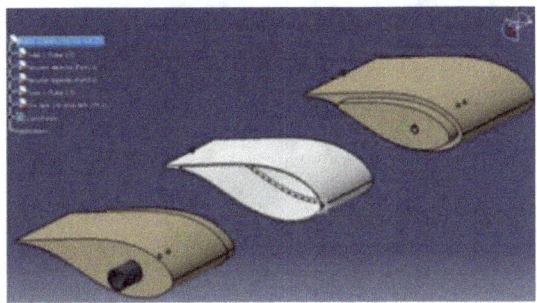

64 inlets pressure:

Scanivalve ZOC33/64PxX2

All these inlets are connected to:

Unit Base of RAD 3200 with 8 modules

Via USB, Scanivalve to PC:

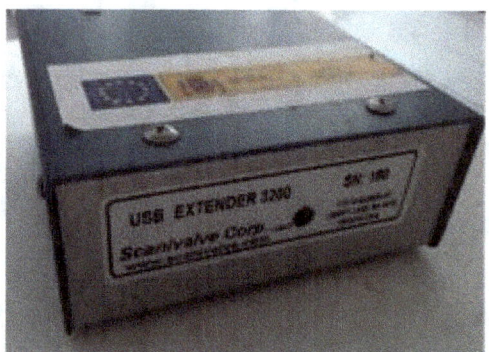

USB Extender.

Is necessary to commute the two modules of 32 connections from:

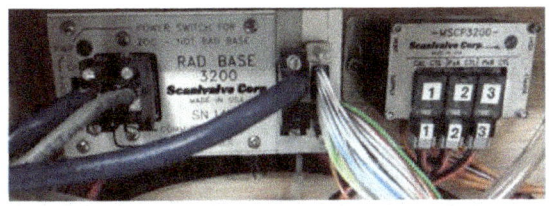

MSCP 3200, right

In order to know the wind speed: Pitot tube:

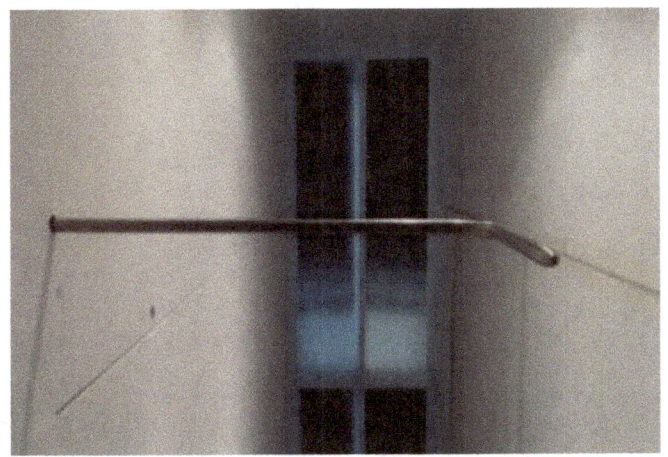

3

DOWNFORCE CALCULATION PROCESS: SECOND METHOD:

Calculate the downforce from force sensor placed in connection between actuator and wing:

It will be necessary to subtract the inertia mass of sprung mass; for that it will calculate the acceleration vertical from next sensor:

Accelerometer.

Advantages:
With second method, is not possible to calculate the drag and moments.
The second method is very simple, easy and cheaper.

4
CALCULATING DRAG:

Consider a particular case where we calculate the drag for an Audi A6: the method can be extrapolated to any other car. To do this, we will put the car at a certain speed and clutch it calculating the time it takes to reach another speed; in this case: from 90 km/h to 70 km/h:
The time it takes to decelerate from 90 km/h at 70 km/h is due to aerodynamic drag.
For this example we have the following parameters:

- ρ_air = 1,225kg / m3 (at 1 atm and 288ºK).
- Incremental factor of mass: γm = 1.04 + 0.0025 · Ɛj² → γm = 1.09
- Deceleration is "equal" to drag.

Therefore:

$$F_r = \frac{1}{2}\rho C_x A V^2 + f_r P = \gamma_m m a$$

$$a = \frac{\frac{90-70}{t}}{3.6} = \frac{5.5556}{t}\frac{m}{s}$$

With:

$$V_m = \frac{\frac{90-70}{2}}{3.6} = 22.222\frac{m}{s}$$

With:
$$m = 1625\ kg; C_x = 0.29; A = 2.26\ m^2; f_r = 0.015$$
Solving:

$$a = 0.2469\ ^m/_{s^2}$$

And:

$$t = 22.5s$$

Reducing the coefficient of drag at 0.03:

$$C_x = 0.26$$

Solving:

$$t = 23.7s$$

We can be changing the coefficient of resistance, until the time actually done.

POLAR GRAPHIC

It exist another form to express the lift and drag of any object, named polar; this graphic is:

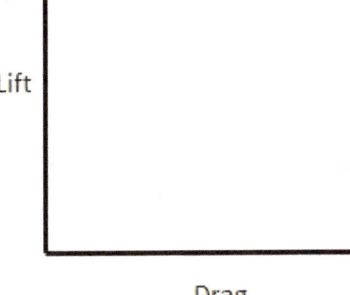

From this graphic, is possible to know a lot characteristics of body tested:

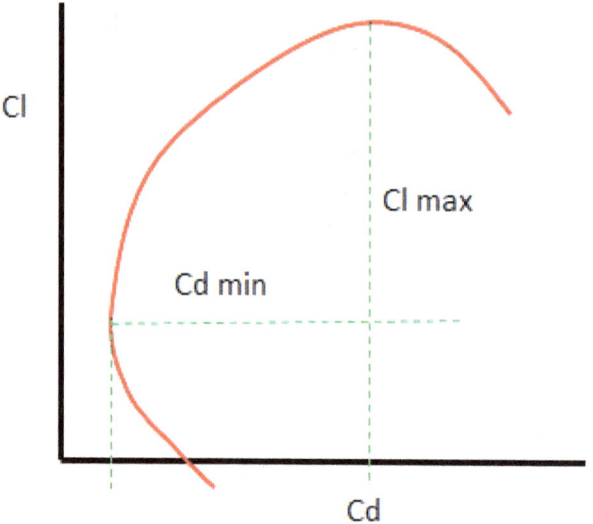

DRAG FRICTION TIRES

The force tractive F_T is:

$$F_T = F_D + M_e \frac{dv}{dt}$$

F_D, drag force full.
M_e, inertial forces (wheels, engine, gear box, etc....).

$$M_e = M + \frac{4 I_w}{R_r^2} + \frac{I_e G_{fd} G_x}{R_r^2}$$

I_w , Wheel inertia.
R_r , Rolling radius.
I_e , Engine inertia.
G_{fd} , Final drive ratio.
G_x , Selected gear ratio.

$$\text{Engine Power} = \frac{F_T.v}{\eta}$$

η - Efficiency of Transmission

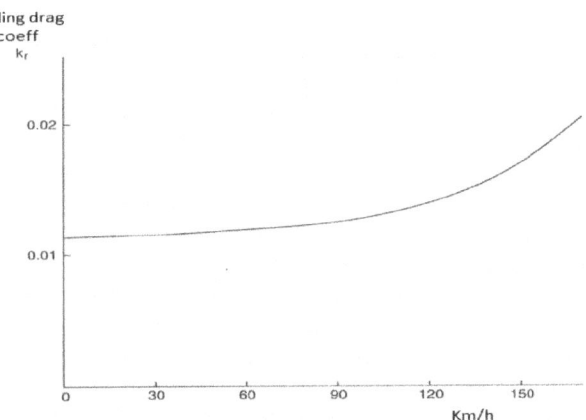

$A_d = 0.015$
$B_d = 8.75 \ 10^{-5}$
$C_d = 0.98$

$$\text{Rolling Resistance} = m.g\left(A_d + B_d v\right)$$

$$Rolling-drag = \left(mg - \frac{1}{2}\rho A\,C_L V^2\right)\left(A_d + B_d V\right)\left(K(1-(T_{std}-T_{amb}))\right)$$

K_T = Tire temperature correction factor (0.011/K)
T_{amb} = Temperature ambient.
T_{std} = Standar temperature.

$$F_{D.Total} = \left(mg - \frac{1}{2}\rho A C_L v^2\right)\left(A_d + B_d v\right)K_c + \frac{1}{2}\rho A C_D v^2 + mg.\sin\theta$$

Let a yaw angle:

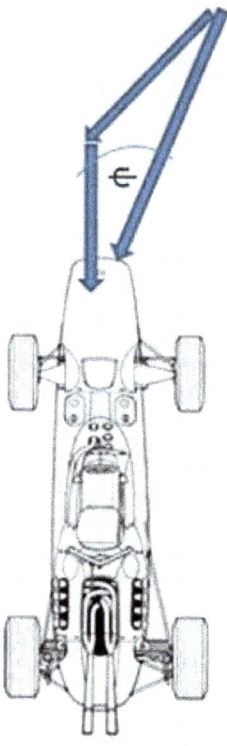

The C_d , depend of this angle: $C_d = C_d(\psi)$:

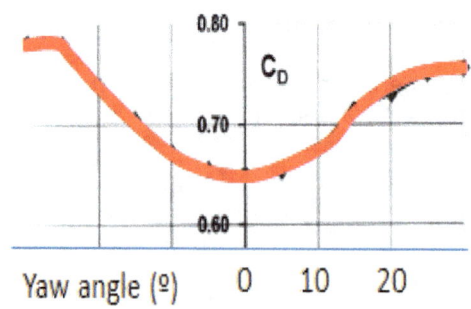

For example, for Ford Scort:

Yaw Angle	Drag Coefficient	Lift Coefficient
degrees	C_D	C_L
0	0.331	0.195
5	0.346	0.224
10	0.375	0.296
15	0.397	0.453
20	0.412	0.618
25	0.427	0.748
30	0.418	0.897

So:

$$F_{D.Total} = \left[\left(mg - \frac{1}{2}\rho A C_L v_r^2 \right)(A_j + B_j v)K_c + \frac{1}{2}\rho A C_D(\psi)v_r^2 \right] + (mg.\sin\theta)$$

→ Is incredible but a lot of thinks, we say about little's incidence angles of front or rear wing (0.1º or som think as that) and in corner, the variation angle of wheel, is VERY GREATER.... And that, have a lot influence in full aerodynamic behavior.

POWER CONSUM BY AERO, ROLLING RESISTENCE PLUS TRANSMISSION

1°) POWER CONSUMED BY AERO + ROLLING RESISTENCE + TRANSMISSION =

a.) 1998 CAR:

FROM WIND TUNNEL TESTS WE HAVE:

DRAG FORCE = 500 (lb)

DOWN FORCE = 700 (lb) } @ 150 (mph) = 241,35 (km)

a₁) - AERO FORCES @ 127,2 (mph):

@ 127,2 (mph) = 204,66 (km/h)

DRAG FORCE: $F_D = 500 \cdot \left(\frac{204.66}{241.35}\right)^2 = 359,53$ (lb)

$$\Rightarrow \boxed{F_D = 1,601.25 \ (N)}$$

DOWN FORCE: $F_L = 700 \cdot \left(\frac{204.66}{241.35}\right)^2 = 503,35$ (lb)

$$\Rightarrow \boxed{F_L = 2,241.79 \ (N)}$$

a-2) POWER CONSUMED BY AERO DRAG:

$$P_D = \frac{F_D(N) \cdot V(km/h)}{(2,654.1)} = \frac{(1,601.25) \cdot (204.66)}{(2,654.1)} =$$

$$\boxed{P_D = 123.47 \ (HP)}$$

a-3) ROLLING RESISTANCE COEFF.

$$f_R = R_C \cdot \left\{ (4.1) \cdot 10^{-3} + (4.1) \cdot 10^{-5} \cdot V \right\} =$$

WHERE:

$R_C = 1.5$ (FOR TARMAC / RADIAL RACING TYRES)

$V = $ CAR SPEED (km/h)

$$\Rightarrow f_R = (1.5) \cdot \left\{ (4.1) \cdot 10^{-3} + (4.1) \cdot 10^{-5} \cdot (204.66) \right\} =$$

$$\boxed{f_R = .0187} \checkmark$$

a-4) ROLLING RESISTANCE : (NOTE TRANSMISSION POWER CONSUMPTION IS INCLUDED IN THE FORMULA)

$$R_R = f_R \cdot (W + F_L)$$

WHERE:

$W = $ TOTAL WEIGHT OF THE CAR $= 1075$ (kg) $= 10,546$ (N)

$F_L = $ DOWN FORCE $= 2,241.79$ (N) ✓

$$\Rightarrow R_R = (.0187) \cdot (10,546 + 2,241.79) =$$

$$\boxed{R_R = 239.13 \text{ (N)}} \checkmark$$

a-5) POWER CONSUMED BY ROLLING RESISTANCE:

$$\boxed{P_{RR} = \frac{R_R (N) \cdot V (km/h)}{(2,654.1)} = \frac{(239.13) \cdot (204.66)}{(2,654.1)} = 18.44 \text{ (Hp)}}$$
✓

a-6) TOTAL CONSUMED POWER @ 127.2 (MPh):
$$\llcorner 204.66 \ (Km/h)$$

$$P_T = P_D + P_{RR} \quad (Hp)$$

$$\boxed{P_T} = 123.47 + 18.44 = \boxed{141.91 \ (Hp)}$$

b) 1993 CAR, BUT RUNNING @ 200 (km/h)

b-1) AERO FORCES @ 200 (km/h):

DRAG FORCE = F_d = 500 $\cdot \left(\dfrac{200}{241.35}\right)^2$ = 343.35 lbf

$\Rightarrow \boxed{F_d = 1,529.2 \ (N)}$ ✓

DOWN FORCE = F_L = 700 $\left(\dfrac{200}{241.35}\right)^2$ = 480.69 (lbf)

$\Rightarrow \boxed{F_L = 2,141 \ (N)}$ ✓

b-2) POWER CONSUMED BY AERO DRAG

$\boxed{P_d} = \dfrac{(1,529.2) \cdot 200}{(2,654.1)} = \boxed{115.23 \ (hp)}$ ✓

b-3) ROLLING RESISTANCE COEFF.

$f_R = R_C \cdot \left[(4.1) \cdot 10^{-3} + (4.1) \cdot 10^{-6} \cdot V \right] =$

$= (1.5) \cdot \left[(4.1) \cdot 10^{-3} + (4.1) \cdot 10^{-6} (200) \right] =$

$\boxed{f_R = .01845}$ ✓

b-4) ROLLING RESISTANCE:

$R_L = f_R \cdot (W + F_L) = (.01845) \cdot (10,546 + 2,291.75)$

$R_L = 236.93 \ (N)$

6-5) POWER CONSUMED BY ROLLING RESISTANCE:

$$P_{RR} = \frac{(235.93) \cdot 200}{(2,654.1)} = \boxed{17.78 \ (HP)}$$

6-6) TOTAL CONSUMED POWER @ 200 (k/h):

$$P_T = P_D + P_{RR} = 115 + 17.78 =$$

$$\boxed{P_T = 132.78 \ (HP)}$$

6-7) DIFF. WITH COAST DOWN TEST:

@ 200 (k/h), TOTAL POWER REQUIRED = 133 (HP)

$$\boxed{\Delta} = 133 - (132.78) = .22 \ (HP)$$

$$\Rightarrow \boxed{\Delta \%} = \left(\frac{.22}{133}\right) \cdot 100 = \boxed{.16 \ (\%)}$$

NOTE:
SEE PAGES 6 & 7 TO COMPARE
RESULTS BETWEEN CALCULATIONS
AND COAST DOWN TEST.

The next image is the comparative between experimental values and numeric calculation:

1998 WILLIAMS RENAULT LAGUNA
Coastdown Test - 26.08.1998

Chassis No: WL98 - 04
Driver: J. Plato
Vehicle Spec: 1998 Aerodynamics Kit, Arch [Spec 1] Sawayguard

Laden Weight - Kgs: 1128

Coastdown Test Equation $F = a + bV^2$

$a =$ 251.5
$b =$ 0.03827

Speeds Removed From Analysis

	kph
CDLS 11	120.0
CDLS 13	120.0
CDLS 14	120.0
CDHS 13	135.0
CDHS 13	105.0
CDHS 14	120.0

Rolling Resistance = $fr = m*g$ $fr =$ 0.02289   Coefficient Of Rolling Resistance

Aerodynamic Drag = $0.5 * Air\ Density * Cd * A * V^2$ $Cd =$ 0.56852   Coefficient Of Aerodynamic Drag

Air Density - kg/m^3 = 1.89
Frontal Area - m^2 = 2.0
Gravitation - m/sec^2 = 9.81

Aerodynamic Drag - Watts = $A*r*V$ V in m/sec
Aerodynamic Drag - kW = $Pir * V / (3.6 * 3.6^3 * 10^3)$ V in kph

1 Horsepower = 746.70 Watts

Vehicle Speed kph	Aero Test Power Consumed Watts	Aero Test Power Required bhp	Rolling Resistance Test Power Consumed Watts	Rolling Resistance Test Power Required bhp	Total Resistance Power Required bhp
0.0	0	0	0	0	0
10.0	11	0	699	1	1
20.0	85	0	1397	2	2
30.0	287	0	2096	3	3
40.0	680	1	2794	4	5
50.0	1329	2	3493	5	6
60.0	2290	3	4190	6	9
70.0	3646	5	4890	7	11
80.0	5443	7	5589	8	15
90.0	7750	10	6288	9	19
100.0	10631	14	6986	10	24
110.0	14148	19	7685	11	29
120.0	18370	25	8383	12	36
130.0	23355	31	9082	13	43
140.0	29170	39	9781	14	52
150.0	35878	48	10479	15	62
160.0	43543	58	11178	16	73
170.0	52228	70	11876	17	86
180.0	61967	83	12575	18	100
190.0	73915	98	13274	19	116
200.0	85944	114	13972	20	133
210.0	99450	132	14671	21	152
220.0	113164	152	15369	22	172
230.0	129342	173	16068	23	195
240.0	145957	197	16767	24	220
250.0	163102	223	17465	25	246
260.0	180841	251	18164	26	275
270.0	200241	281	18863	27	306
280.0	233362	313	19561	28	339
290.0	256269	348	20260		375
300.0	287025	385	20958		413

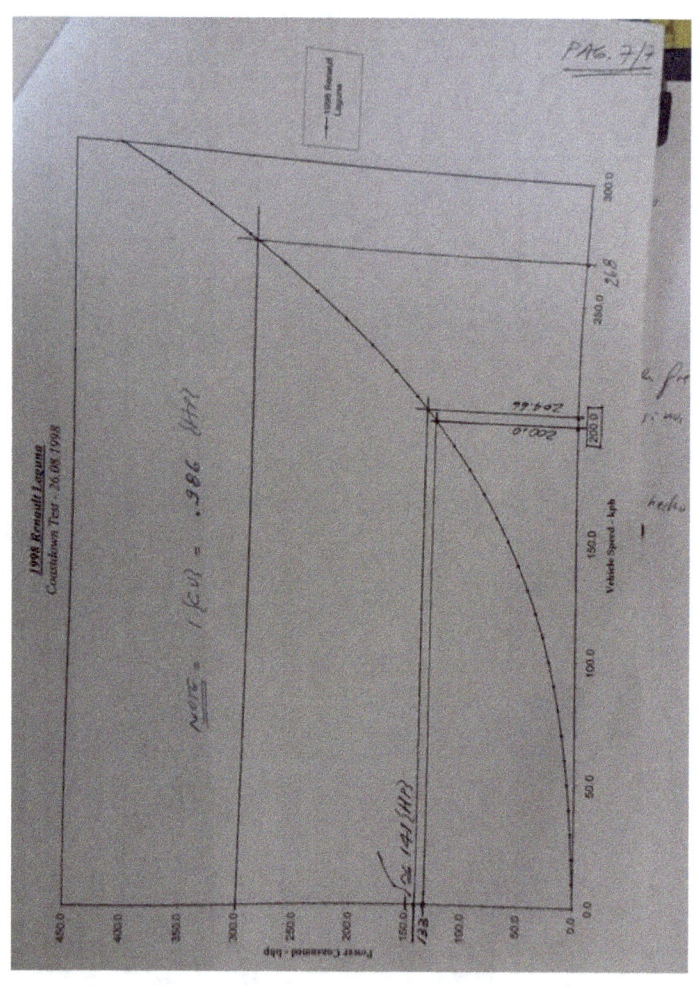

Example:
San Diego State University
Aerospace Engineering
AE 403 – Senior Project
<u>Aerodynamics of a Modified Racecar</u>
Hoang Pham
James Knerr
Khan Mohammad
Professor Joseph Katz

Table of Contents

Abstract

Performance analysis through wind tunnel testing was performed over a modified racecar. Non-dimensional coefficients of lift and drag were experimented with variations in aerodynamic lift devices. A computer-aided design of the vehicle was done through Pro/E (Creo 5.0) and an 18% scale was created for testing in the San Diego State University wind tunnel. In general, cars have an elementary shape that is in the form of an inverted airfoil. The clean configuration of the car had an undesirable positive lift, which made the vehicle unstable when traveling at high velocities. Therefore, the objective was to gain a zero lift coefficient on the vehicle to fix the threshold of the dependence of drag on lift. Then various configurations were analyzed to see which one would produce the lowest drag coefficient possible. **Introduction** The objective of this experiment was to achieve aerodynamic forces that would favor the vehicles objective to break the land speed record for a pickup truck. Aerodynamic forces increases significantly as velocity increases. Drag forces will reduce the top speed due to skin and pressure friction. Furthermore, because the car is shaped in an elementary form of an airfoil, positive lift (up from ground) could be catastrophic. Although a high magnitude of L/D can be achieved, as lift is increased drag also increases. Down force will not affect the cars velocity, as it is perpendicular to the velocity vector and will contribute no work. Therefore, a zero coefficient is most effective because this will yield the desirable stability results while providing the lowest coefficient of drag for the vehicle. The lowest coefficient of drag will also benefit the vehicle when reaching its terminal velocity.

Theory In order to qualitatively analyze experimental data from a scaled model to that of the life scale, non-dimensional coefficients were determined with respect to the frontal projected area of the car. In doing so, valuable data could be correlated between the two different scales. The scaling of the model was done with help of Professor Katz. His experiences helped determine a large enough model where force readings would be more accurate. At the same time, the boundary layer of the wind tunnel would not come into effect when doing the analysis. In addition, application of fluid mechanics to convert wind tunnel readings and aerodynamic equations are provided next.

Nomenclatures:

L	Lift force
D	Drag force
q	Dynamic pressure
S	Model frontal area
v	Velocity
C	Wind tunnel reference area
ρ	Density of fluid medium
C_L	Coefficient of Lift
C_D	Coefficient of Drag

Figure 1.1

Equations:

$$C_L = \frac{L}{\frac{1}{2}\rho v^2 S}$$

$$C_D = \frac{D}{\frac{1}{2}\rho v^2 S}$$

$$gh_a + \frac{P_a}{\rho} + \frac{u_a^2}{2} = gh_b + \frac{P_b}{\rho} + \frac{u_b^2}{2}$$

$$q = P_{total} - P_{Static}$$

Experiment Procedure:

After receiving the geometry from Mike Peters, completion of the CAD model was performed on PRO E 5.0 with an 18% scale. The file was converted to a Para solid then transferred to MASTERCAM to be finished into a surface that can be machined. MDF wood was glued together in layers to achieve a slightly bigger geometry then was expected.

Then a three-axis CNC machine was used to mill out the model with the MASTERCAM program. The model was then sanded, prepared, and painted with automotive paint. To replicate the skin friction of a real car two coats of wax were used prior to wind tunnel testing. Pictures of the model and various aerodynamic devices are presented in Appendix A.

The front splitter, spoiler, rear diffuser, and winglets were all made with geometry proportional to that seen in historical data and with calibration with Professor Katz. The change in front splitter length and rear deflection angle can be seen from a mock up picture provided in Appendix A.

At the university, the wind tunnel was modified to have a ground plate. Four rods protruded from the balance underneath without touching the ground plate to under the car. Then the car was positioned so that the wheels were off the ground plate by three mm. Aluminum tape was used in conjunction with foam to stimulate a ground effect in front of the wheels.

This was done so that the balance could yield purely the reading of the forces without any physical interference with the wind tunnel. Details of this are presented in Appendix A as pictures. Fluid properties and dynamics were calculated from properties that were recording during the time of experiment from **Figure 1.1**.

	Day 1	Day 2
Hg (mm)	29.9	29.6
Temperature (F)	61	76.5
Dynamic Pressure	3.5	3.5

Figure 1.2

The H_2O level throughout the test had a 1/16inch deviation during testing and was neglected in the analysis. At the end of the experiments, the barometer and temperature reading remained the same.

Furthermore, each reading was zeroed by an empty run with only the four rods, nuts and ground plate in the wind tunnel. This was done to calibrate the model inside the wind tunnel. Tests were done at various configurations with data output in pounds of lift and drag. For the readings for each run please see Appendix B.

Results

All of the various components of each configuration are considered fixed except the front diffuser and rear spoiler deflection angle. The data consist of graphs with varying components, splitter length, and spoiler deflection angle. Negative lift forces the car into the ground and positive drag forces the car backwards.

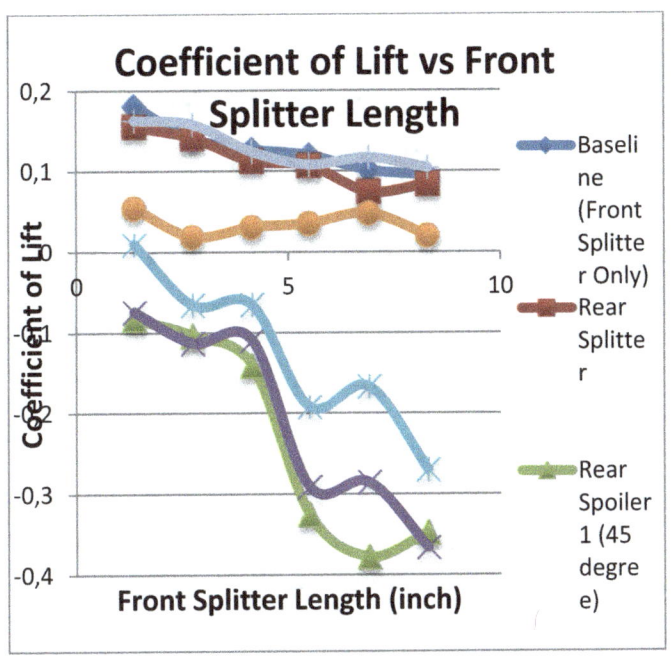

Figure 1.3

From **Figure 1.3** the coefficient of lift decreases as the front splitter length is increased. Especially in the case of the three spoilers, the coefficient of lift decreases rapidly as splitter length is increased. Notably, the spoiler deflection angle also plays a crucial role in the increase in negative lift. The rear splitter and cabin flaring, when compared to the baseline shows no notable changes as the front splitter length is increased. Therefore in this configuration of two devices, it is assumed that the rear splitter and cabin-flaring coefficient of lift is independent of front splitter length. Furthermore, the coefficients of lift for the winglets are independent of front splitter length change as well.

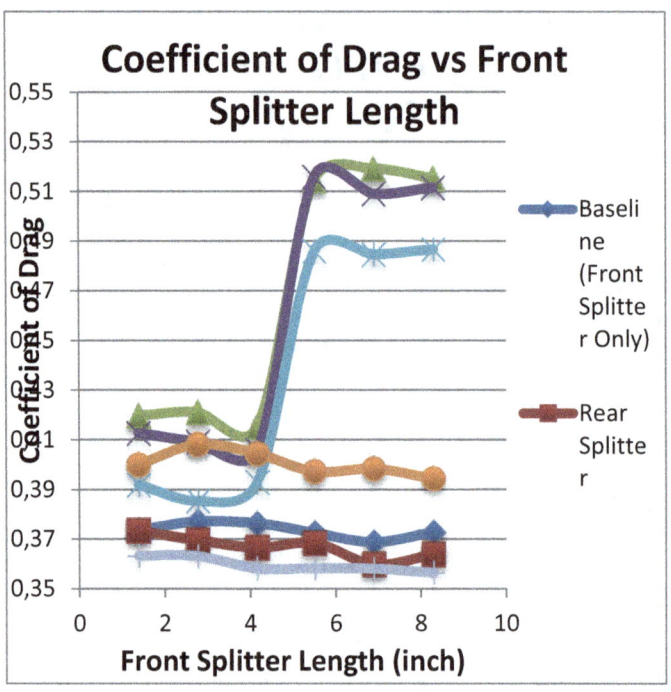

Figure 1.4

From **Figure 1.4** the overall coefficient of drag for the three spoilers significantly increase with increasing front splitter length. This may be due to the contribution of streamlines that are directed over the car to the spoilers. The winglets although produce desirable lift, their drag coefficient at times match's the induced drag seen at the spoiler. This was seen troublesome as there were other spoilers that produced more lift but without as much drag.

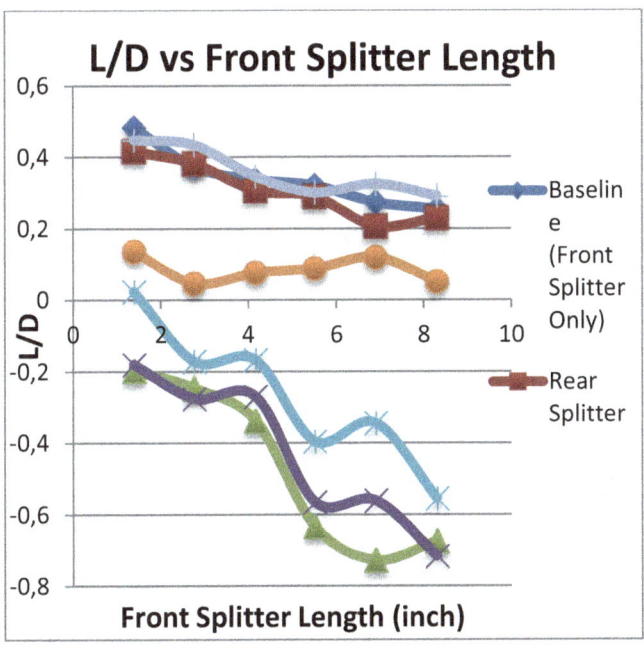

Figure 1.5

Figure 1.5 shows the coefficient of lift and drag for the various components. To note, the rear spoiler 3 with 19-degree deflection angle crosses the x-axis to provide zero lift with a particular front splitter length. From **Figure 1.4** the coefficient of drag was slightly over .39 for the same configuration. Our final configuration was different in order to reduce drag we had to use the rear splitter, which in turn when compared to this configuration gave us a lower coefficient of drag.

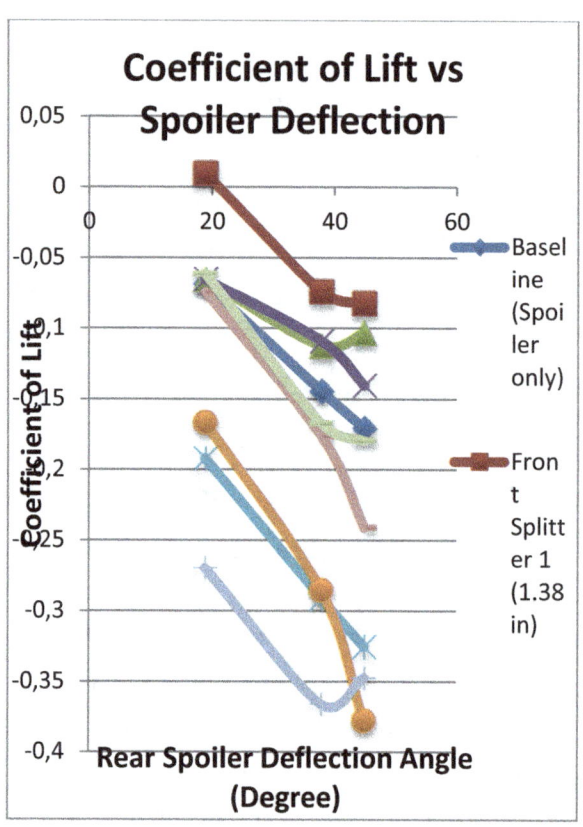

Figure 1.6

The coefficient of lift was graphed with changing rear spoiler deflection angles on the various components of the car. The most desirable results from the lowest rear spoiler angle with the shortest front splitter length. At the point where the Front splitter 1 (1.38in) crosses the x-axis gives us the minimum drag with zeroed lift. The drag coefficient can be seen in **Figure 1.7**. When the spoiler 1,2 and 3 were used in conjunction with front splitter 1,2, and 3 the results yielded a coefficient of lift closer to zero. The higher deflection angles produced a high negative lift but this will be accompanied with an increase in induced drag.

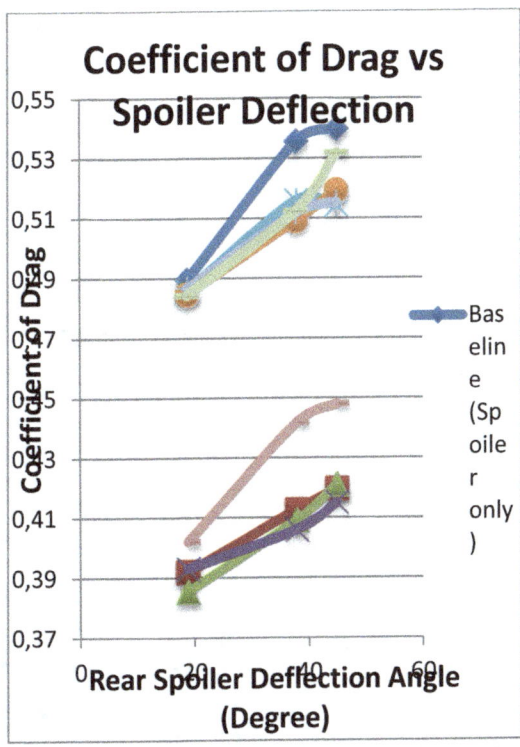

Figure 1.7

In **Figure 1.7**, the coefficient of drag with rear spoiler deflection angle was graphed. The baseline figure for the spoilers alone yielded the highest coefficient of drag. However, when used in conjunction with other aerodynamic devices, the overall coefficient of drag was reduced. The rear spoiler's coefficient of drag suffers due to induced drag. However if we note that the rear splitter when used in conjunction with the spoilers produces the lowest drag. From Reference [1], the rear splitter in this case creates a high pressure-trailing vortex that would reduce the pressure gradient from the front of the car. Thereby, reducing drag even with the rear spoilers attached. In addition, as the front splitter length is increased, the coefficient of drag rapidly increases. With increasing front splitter length, the streamlines that travel over the car are increased. These increasing streamlines produce an increase in net fluid flow over the rear spoiler.

Therefore, it would also increase the induce drag caused by the spoilers. The same picture can be said in **Figure 1.4** as splitter length is increased.

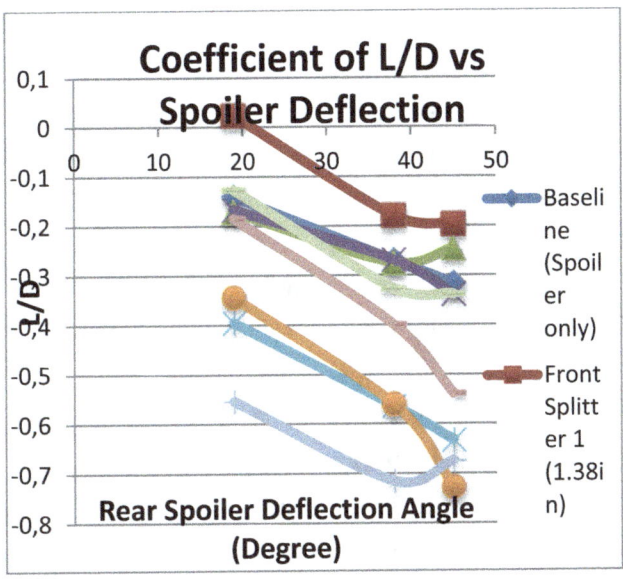

Figure 1.8

The coefficient of lift over drag is presented in **Figure 1.8** above and shows more desirable results can be obtained from a shorter front splitter length and rear spoiler deflection angle over the baseline of just the rear spoilers alone.

Conclusion In theory, the fluid flow behavior behind a car affects the fluid flow ahead and vise versa. This complicates the subject because no single data from these tests can be super imposed to demonstrate the entire fluid flow behavior around the car. Testing was done systematically to observe general behavior of aerodynamic devices. Afterwards, with some reasoning, different configurations were formulated to produce a desirable result. With a zeroed lift, there is still a net force that produces negative lift as compared to the original configuration.

Therefore, if the car was to bounce over an incremental bump, there is a net negative lift forcing the car to "stick to the ground." The car would still benefit from a stability aspect at zeroed lift. With this in mind the zeroed lift would give a threshold to the coefficient of drag.

Although the rear spoilers and winglets produced very high drag coefficients, in order to produce a lift near zero the smallest deflection angle spoiler was used. As previously mentioned when the spoiler was used in conjunction with the front splitter, the coefficient of drag was seen to decrease. The smallest rear spoiler angle and shortest front splitter length was used. After concluding this, many configurations were experimented with. As a result, the configuration (f0,s1,rs) produced a -0.0000823 coefficient of lift with a 0.3873 coefficient of drag. From the lowest drag of the experiment of 0.346354, the percent different between these two values are 10.57%.

Configuration	C_L	C_D
Front Splitter @ 8.3 in Cabin Flaring Rear Splitter	0.0971017	0.34285
Front Splitter @ 0.694 in Rear Spoiler @19 degree Rear Splitter	-0.000823	0.3873

Figure 1.9

- Although the second configuration produced the lowest net drag, the net coefficient of lift was still positive.
- The rear spoilers were plagued with induced drag, however spoiler 3 was necessary to produce a negative but closer to zero lift.

Design Recommendations

Future design recommendations would be to test an inverted airfoil to produce down force. Since airfoils are engineered to have the least drag while producing high lift, we can minimize the coefficient of drag that accompanies the coefficient of lift.

References

[1] Katz, J. Aerodynamics of Race Cars, The Annual Review of Fluid Mechanics, 2006. 38:27-63